Dear Jane

Congratulations in all you are doing. With my blessings.

David Ash

26/02/25.

Was Einstein Right?

David Ash

Other books by David Ash

The Vortex: Key to Future Science, (with Peter Hewitt) Gateway Books, 1990

The Science of Ascension, Cloverleaf Connections, 1994

The New Science of the Spirit, The College of Psychic Studies, 1995

Activation for Ascension, Kima Global Publishers, 1995

The New Physics of Consciousness, Kima Global Publishers, 2007

The Role of Evil in Human Evolution, Kima Global Publishers, 2007

The Vortex Theory, Kima Global Publishers, 2015

Continuous Living, Kima Global Publishers, 2015

AWAKEN, Kima Global Publishers, 2018

SUPERENERGY and the Quantum Vortex, Mill House Publishers, 2023

"A theory is more impressive the greater the simplicity of its premises, the more different things it relates, and the more expanded its area of applicability."

- Albert Einstein, 1949

Was Einstein Right?

10 ESSAYS
BY
DAVID ASH

Images by
SHANE GUFFOGG

Hardcover book, English edition

ISBN 978-87-94302-85-2

Art by Shane Guffog © 2025

Edited by Dawn Austin Locke

Published in Denmark

Mill House Publishers

https://www.millhouse-publishers.com

Contents

To my father,

who engendered in me a love of physics.

Acknowledgments

Catrin Platt, thank you for accepting these essays for the Philosophy Society of Exeter University. John Cliss and Oliver Percival, thank you for your support of my endeavours over many years. Ashen Page and Matthew Newsome, I thank you for encouraging me to write this book and your enthusiasm for what I am doing. Lorna, my youngest daughter, and Ru, thank you for your love and support in my endeavours. Lorna and Michelle, thank you for introducing me to Substack as an online platform for review of the essays. Sam, my son, thank you for encouraging me to keep writing. Annemarie, I thank you for your belief in my work and for publishing my books and laying them out so beautifully. Most especially thank you for taking on this book and arranging for it to be translated into Danish. Chrissy, Tara and Neil, thank you for interviewing me on your broadcasts which has helped me bring my work into focus. I have Walter Isaacson to acknowledge for his amazing book on Einstein that inspired the title and theme of this book.

I have my father to thank for engendering in me a love of physics and setting me on the path of original thinking in science. Though he has passed on, he is still an inspiration for all my endeavours. My mother also, I thank her for caring for me when I was supposed to be caring for her and for the effort she made at age ninety-five to understand Heisenberg's uncertainty principle. Anna, my first wife, I thank you for your lovely drawings of hedgehogs and tadpoles as well as elephants and ants in their bubbles of space, Jessica, my second daughter, thank you for your delightful picture of Democritus, and Rebecca, my third daughter, thank you for your charming drawing of Mr Proton. Josephine, my eldest, thank you for your drawings; and Ondine, Lulu and Lily thank you also for being such loving daughters. I thank my brother Steven and his wife Renata for their love and support of all my enterprises over many years. Also, my sisters Jenny and Mary and brothers Simon, Peter and Richard for their love and encouragement over many years.

I would like to express my heartfelt gratitude to Shane Guffogg for including his amazing artwork in my book, and to Phillipe and Linda Warin for their hospitality to Suzie and me in Paris and inviting us to the opening of Shane's exhibition there. And I wish to thank Simon Best, editor of Caduceus magazine, for his encouragement and helpful advice especially in regard to the figure captions in this book. Most of all, I have Suzie Thompson to thank for her ongoing love and support. Not only has she been constant in helping me write and edit the essays but in careful listening and discussing the physics she made a significant contribution, bringing my attention to the need in quantum mechanics for a principle to define what is and what is not a real particle. In acknowledgement of this I put her name to it.

Introduction

Presented as a series of ten essays, written for the Exeter University Philosophy Society, this book represents the culmination of over fifty years of dedicated research. It begins with Einstein's challenge to the uncertainty principle of Werner Heisenberg. Einstein was right because, five years after the uncertainty principle was published, a nuclear particle was discovered that proved Heisenberg's principle was wrong. Heisenberg immediately objected. The physics community should have exonerated Einstein. Instead, they continued to back Heisenberg.

For the best part of a century the world physics community hasve ignored the evidence about the structure of the nuclear particle in order to save the uncertainty principle and the quantum mechanics that stemmed from it. The word ignorance is derived from the verb to ignore. If scientific evidence is ignored in order to shore up incredulous theories then science is in danger of embracing ignorance, in the true sense of the word.

I have written this book because I believe Einstein was right in his stand against Heisenberg. When I discovered Heisenberg's theory had been disproved, the evidence covered up and Einstein has been discredited, I decided to blow the whistle. Theories should be ditched if they are disproved by experiments. That is scientific law. As quantum mechanics is based on the uncertainty principle it needs to be ditched or restructured around a sound principle.

The quantum mechanics developed from Heisenberg's uncertainty principle is complex and incomprehensible. Richard Feynman remarked that nobody understands quantum mechanics, they just get used to it. That suggested to me there is something fundamentally wrong with it. In my teens I discovered a principle that could replace uncertainty as the underpin of quantum mechanics. The discovery enabled me to develop a new approach to quantum physics that is simple, comprehensive and understandable, which also points to the possibility of a credible alternative to the principle of uncertainty at the core for quantum mechanics.

The discovery I made when I was sixteen enabled me to restructure quantum mechanics, through art rather than math, and present it in plain language. I used analogies rather than equations so that, with minimum knowledge of the subject, anyone can appreciate it and be enchanted by it. It has taken me half a century to distil the ongoing realisations from that initial discovery and develop them with new insights and different means of expression. I can now deliver a clear and simple account of quantum mechanics restructured, one that is knowable, realistic, compatible with the rest of physics, and congruent with the structure of the known world.

As the father of quantum theory, Einstein had a responsibility to challenge the architects of quantum mechanics when they determined the quantum world was unknowable. Who else could insist it is essentially knowable? Einstein took a stand for realism when realism was dismissed as naïve. Einstein's stand for realism was vital for me because what I had stumbled on in my youth was a visualisation of quantum spin and it was their inability to visualise quantum spin that caused quantum theorists to despair of realism. My revelation determined that the spin of energy is the basis of matter. So it not only revealed that quantum reality can be known in a very realistic way, it also provided a vivid image of how energy forms mass as predicted by Einstein. The implications of my discovery were immense. It came from an ancient lost knowledge that anticipated Einstein by millennia.

On my journey, Einstein was the guiding light who helped me on my way. As I delved deeply into his life and work, I imagined him as a companion. Often, when I wrestled with a problem and was on the point of giving up, a simple and obvious solution would come to me. Einstein had the same experience in his musings as he paced many a room in cosmic contemplation, twiddling a forelock of his hair.

We have a well of knowledge inside us that we can delve into for answers to any questions we ask. It is an inner knowing that connects us to a universal source of inspiration. I could never have done what I have had I not been aided by innate knowledge and inspiration. So much did I depend on it that I came to realise my job was not to find the right answers but to ask the right questions. The answers would come. That was Einstein's experience also. It was from his inner well of knowledge he knew intuitively that Heisenberg was wrong.

The ten essays reveal the value of intuition as well as reason in science. They not only outline a new underpin for quantum mechanics, they also round off physics and show how it is possible to approach the subject in a non-mathematical way. Rather than add formulae to physics the essays focus on fundamental understandings. They include philosophy because quantum physics has profound philosophical implications. The unified physics they present is based on a single insight that has far reaching implications beyond science. They speak for the future as well as the past. Most important of all, the essays are simple, self-consistent and easy to comprehend – and they embrace art.

As soon as I saw Shane Guffogg's work I was immediately captivated and invited him to include his art in my book, especially as he spoke of his work looking through the lens of humanity at civilisations past and present, viewing time as threads that connect all people. Never before had I seen such a vivid depiction in art of the insight I work with. A lost knowledge from an ancient civilisation, it is a thread spinning through time that has the potential in our present day to unify not just physics but also humanity.

1. Neither Flesh nor Fleshless - Shane Guffogg

*I chose this image because for me it beautifully de-
picts quatum waves of energy* - Ash

ESSAY 1

Uncertainty about Uncertainty

"If your theories and mathematics do not match up to the experiments then they are wrong."

- Richard Feynman

Précis: In 1919 Albert Einstein became the most celebrated scientist in the world. Ten years later he was a has-been. Despite his achievements he was no longer taken seriously by other physicists. This was because he denounced the uncertainty principle of Werner Heisenberg and the quantum mechanics coming out of the centre for theoretical physics in Copenhagen. Einstein was a realist. He didn't go along with the idea that the quantum world is unknowable and the Universe depends on our observation of it for us to be certain of its existence. At the time Einstein was unable to prove Heisenberg wrong. However, a few years later a discovery was made that disproved the uncertainty principle but, instead of endorsing Einstein, the physics community ignored the evidence and continued to back Heisenberg.

In 1927 Heisenberg published a theory that the more certain you are of the position of a particle the less certain you are of its velocity. In 1932, in Britain, James Chadwick discovered a new nuclear particle; the neutron. It was formed of an electron and a proton – the two particles that had been discovered to make up atoms – it fell apart into an electron and a proton, and its mass was that of an electron and a proton. The neutron appeared to be an electron bound to a proton by opposite charge attraction. It appeared to be neutral because the opposite charges of the two particles cancelled each other out. However, Heisenberg objected to this because if the neutron was an electron docked on a proton you could be certain of its position. That would mean, according to his formula, that it would either be at a standstill or rushing round at nearly the velocity of light. That was impossible.

Though it was immediately evident his theory was wrong, the physics community went along with Heisenberg's objection. They discounted the evidence against the uncertainty principle and the quantum mechanics that stemmed from it and determined to find a way to prove the neutron was a unique chargeless particle. A few decades later that was achieved through a new theory of subatomic particles called quarks. A new force of nature was proclaimed. Called the 'weak nuclear force', it was declared to be capable of transmuting the quarks so that if an electron became attached to a proton it would immediately disappear and the proton would undergo transformation into an entirely chargeless neutron.

Realism and common sense had been abandoned by a group of headstrong young theorists linked to the Copenhagen Institute of Theoretical Physics founded by Neils Bohr. Dismissing realism they lost sight of reality and Einstein was dismissed. They may have been jealous of his celebrity status and might have ignored his misgivings because of his age but it was impetuous of them to turn their backs on the physics that had gone before them and to abandon the single most important principle in science that no matter how brilliant a theory may be, or how well it works mathematically, if it doesn't agree with experiment it is wrong.

Fig 1 Albert Einstein

Essay: Albert Einstein was the greatest scientific genius of modern times. In the space of a single year, before he even had a PhD, he published four papers which established the atomic theory, quantum theory, the special theory of relativity and the groundbreaking equation, $E=mc^2$. He was then 26. By 34 he had a professorship in Berlin and was the youngest member of the prestigious Prussian Academy of Science. Then at 36, with the publication of his general theory of relativity, he appeared to have toppled Newton. When his theory seemed vindicated by the Royal Society, in 1919, he became the most celebrated scientist in the world. At 40 he was awarded the 1921 Nobel Prize in Physics for founding quantum theory. But in the late 1920's, as he approached 50, he plummeted from the pinnacle of physics. He was no longer taken seriously by professional physicists. They regarded him as little more than a remarkable oddity and as he advanced into the last quarter century of his life, despite his tireless work on field theory, he was disregarded as a reactionary old man who was well past his prime.

So, what happened?

In 1927, a young German theorist, Werner Heisenberg, commandeered quantum theory when he published his now famous principle of uncertainty. He initiated a new system that came to depend on it called 'quantum mechanics' which replaced the 'old quantum theory', becoming the fulcrum of physics. Einstein, who originally established quantum theory, despised the uncertainty principle. He described it as, "...a real witches calculus...most ingenious, and adequately protected by its great complexity against being proved wrong." [1]

It was Albert Einstein's outspoken criticism of the uncertainty principle and quantum mechanics that cost him his lead and reputation in physics. He was no longer driving the quantum bandwagon he had set in motion. A Dane had seized the reins from him then handed them over to a young Hun who hurtled off in an uncertain direction with a haggle of headstrong young Turks behind him. Einstein was left in the ditch. The new generation of physicists were not listening to him. In the abandon of youth they were deaf to the older man's pleas for common sense.

Meanwhile, in England an intrepid New Zealander, Ernest Rutherford, had been probing the atom with the mysterious rays of radioactivity. The old professor at Cambridge, Sir J. J. Thompson, had discovered the electron in 1897 and, in 1911, Rutherford uncovered a central nucleus in the atom. He suggested that electrons orbited the nucleus like the planets orbit the Sun, but there was a problem with his solar system model for the atom.

Fig 2 Niels Bohr

Then, in 1913, one of Rutherford's former students, a 28-year-old Danish theorist, Neils Bohr, overcame the problem by applying Einstein's quantum theory to the orbiting electron in a hydrogen atom. He determined that electrons exist in atoms in certain quantum states and that they can take a quantum leap between these states as they emit or absorb energy. He described the atom as we know it today. For his achievement Bohr was awarded the 1922 Nobel Prize. In 1916, after being appointed professor of physics in Copenhagen, Bohr determined to pull the epicentre of quantum physics to Denmark. He began to recruit brains. By 1920 he had assembled a team of brilliant young minds for his new Copenhagen Institute of Theoretical Physics, which formally opened in 1921. A star student in his faculty, Werner Heisenberg, was amongst them.

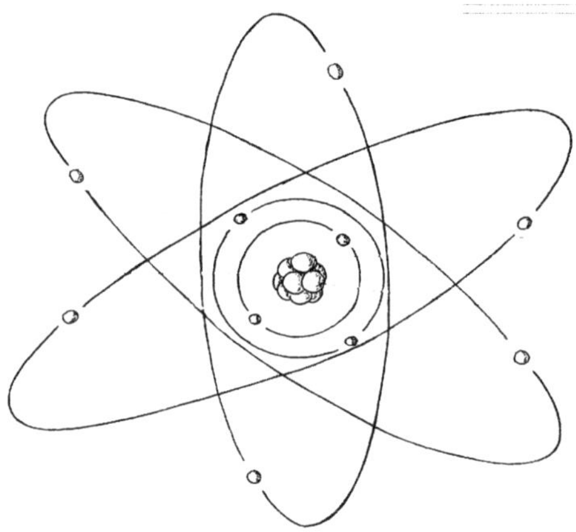

Fig 3 The Atom

One evening in 1922, Heisenberg was strolling in a park pondering on the quantum leap of electrons in atoms. A lamp lighter was lighting the gas street lamps and as the shades of night drew in Heisenberg noticed a man walking along the path underneath them. He could only see the man when he walked into the light of a lamp. When he continued into the dark between the lamps the man disappeared. Heisenberg realised he could only be certain of the man's existence when he interacted with the light of a lamp. In the dark, when the man was between the lamps, Werner was no longer certain he was there. He concluded that likewise, as an observer, he could only be certain of an electron in the atom when it was absorbing or emitting light; otherwise he could not be certain of its existence. That would lead eventually to the Copenhagen interpretation of quantum reality, which concluded we can only be certain that things exist when they are observed.

For most of us it would appear obvious that if the Universe has been around for 13.5 billion years and Homo sapiens appeared within it only 190,000 years ago it must have existed before us; we are, after all, made of its ancient atoms. That common sense is called realism. But in the Copenhagen interpretation that realism was dismissed as naïve. The argument was we could only presume that the Universe existed before we came into existence. We could not be certain because we were not there to observe it. Like the man in the dark walking between the gas lamps, even if it existed, the Universe was in the dark of not yet being observed, at least by humans. Common sense doesn't apply in quantum mechanics. Even Einstein couldn't succeed in bringing the common sense of realism into it. In 1964 a physicist from CERN, John S. Bell, thought he had won the argument for Einstein in his now famous Bell's Theorem but when that failed he despaired of quantum mechanics, "I hesitated to say it was wrong but I knew it was rotten."[2] It is not quantum mechanics that is rotten it is the uncertainty principle that underpins it. The rot set in when it was used to develop quantum mechanics despite being disproved by the discovery of the neutron, which occurred five years after its publication.

In the year Einstein became an international celebrity Rutherford had discovered a positively charged particle in the atomic nucleus, which he called a 'proton'. It was 1836 times as massive as a negatively charged electron. From his study of radioactive decay Rutherford predicted the existence of a neutral particle in the nucleus, which he named a 'neutron', formed from an electron combined with a proton as a result of their mutual charge attraction. In the process of radioactivity, he said, the neutron decayed back into an electron and a proton. Rutherford's maxim was, "These fundamental things have got to be simple", so he surmised the neutron would appear to be neutral due to the opposite charges of its composite electron and proton neutralising each other. In his mind the neutron was an engagement of electron and proton.

Fig 4 Ernest Rutherford

One of Rutherford's students, James Chadwick, sallied forth on a hunt for the neutron and refining some of Rutherford's research techniques, he discovered it in 1932. He was awarded the 1935 Nobel Prize in physics for this achievement. The profile of the neutron fitted Rutherford's prediction. Its mass was about 0.1% greater than a proton. This corresponded to slightly more than the sum of the mass of a proton and an electron. In a process called K-capture neutrons could be formed from an electron captured by a proton in an atomic nucleus and when neutrons decayed in a radioactive nucleus, in a process called beta-decay, they released a proton and an electron. For all intents and purposes the neutron appeared to be a bound state of an electron and a proton, just as Rutherford had proposed.

Fig 5 James Chadwick

Werner Heisenberg was quick to object. He proclaimed the neutron could not be a bound state of an electron and proton. It had to be a unique neutral particle. Why was it necessary for him to suggest that when there was clear experimental evidence supporting the bound state model? Why did the entire world of physics snap into step behind him? Even today physicists will assure you a neutron cannot be an electron docked on a proton. This is because an equation Heisenberg was working on had led him to conclude that the observer effect was operating at a quantum level. The observer effect is that measurements of certain systems cannot be made without the act of observation affecting them. He determined that the process of making certain measurements in the subatomic world would increase the uncertainty about what was going on there. For example, if you wanted to look at a subatomic particle in order to ascertain its position you would reflect a quantum of electromagnetic radiation off it. But the act of bouncing a quantum of energy off a subatomic particle would give it a kick, which would increase its momentum, and that would make its position more uncertain. Looking at small objects requires more energy than looking at large ones. This is evident in the electron microscope, which employs higher frequency radiation than an optic microscope. Because subatomic particles are the smallest things in nature the action of ascertaining their position with any degree of certainty would require the use of very large amounts of energy, which would give them an enormous kick. On the basis of this, Heisenberg argued against physicists being able to determine with certainty both the position and momentum of a subatomic particle. This inferred indeterminacy between time and energy..

Fig 6 Werner Heisenberg

Heisenberg included his ideas of quantum indeterminacy with other features of quantum theory including wave-particle duality determined by Louis de Broglie and wave indeterminacy proposed by Erwin Schrödinger. Together they were incorporated into a formula known as the Heisenberg uncertainty principle, which was central to the Copenhagen Institute's development of quantum mechanics from 1927 onwards. However, when Heisenberg's formula was applied to the Rutherford-Chadwick neutron, as an electron locked onto a proton, it failed. This is because the high degree of certainty of an electron in a neutron suggested such an enormous indeterminacy in its momentum that electrons could be racing round on the surface of a proton with velocities up to 99.97% of the speed of light. Einstein's theory of relativity determined that a moving particle is more massive than a particle at rest. At

99.97% of the velocity of light electrons would possess forty times the mass of an electron at rest. The wide range of electron velocities predicted by the uncertainty principle would be reflected in a measurable range of mass values for neutrons, from that of a single electron and proton to that of a proton and forty electrons. In fact all neutrons have precisely the same mass of 1674.92 x 10–30 kg. Thus the neutron provided an experimental test that proved the uncertainty principle was wrong.

Heisenberg was not willing to concede defeat. He was unwilling to accept that his uncertainty principle had been disproved so he insisted that neutrons are not electrons bound to protons. In defiance of the rules of science he disregarded the evidence presented by the neutron in order to protect his theory and the physics community closed ranks around him because by 1932, when the neutron was discovered, quantum mechanics had gained so much momentum that physicists were reluctant to accept the uncertainty about the uncertainty principle underpinning it. They decided to ignore the evidence about the neutron until they could circumvent it. They achieved that in the 1960s with the invention of quark theory.

Quarks are hypothetical particles attributed with fractional charges. Because fractional charges have never been observed in nature to suggest a proton is made of two up-quarks with ⅔ charge and one down-quark with ⅓ charge, which add up to unitary charge, is a totally arbitrary assumption. The suggestion that a neutron is made of two down-quarks with ⅓ charge and one up-quark ⅔ charge that add up to zero charge reveals quark theory is an attempt to get round the evidence that a neutron is an electron bound to a proton; and the theory is maddeningly complex.

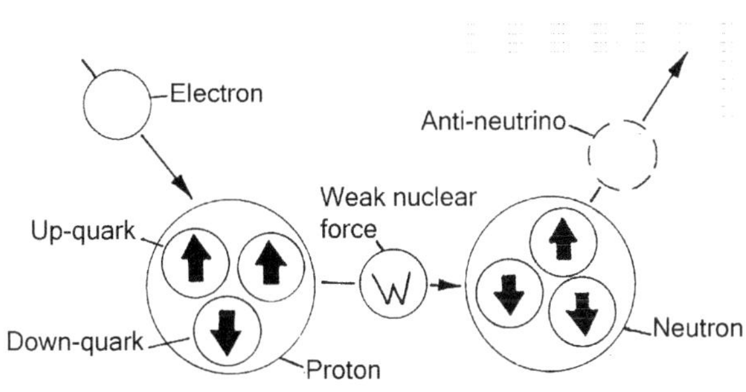

Fig 7 Transmutation of a proton into a neutron

According to quark theory, when an electron interacts with a proton a weak nuclear force comes into play which transforms an up-quark into a down-quark and the electron is temporarily lost. In this process the electron and proton supposedly cease to be and their place is taken by a neutron and an anti-neutrino. When a neutron decays, the down-quark reverts to an up-quark and the proton reappears with the disappearance of the anti-neutrino and appearance of an electron with a neutrino, which carries away some of the electron's energy. These fundamental things are certainly not simple.

The quark account for neutrons reminds me of the story of a mad professor of puddings:

Once upon a time in a crazy world there was a mad professor of puddings. A student in his department of nutrition baked plums in a pudding then ran round shouting, "Eureka, I have just invented plum pudding!"

The professor was not impressed. He snorted over his spectacles, "That is not a plum pudding. That is a Black Forest gateau."

The downcast student argued, "But I mixed plums into my pudding before I baked it and when I weighed it the weight was that of the plums and the pud, and at the end, when I shook the pudding, out came a plum. It has to be a plum pudding!"

The professor became really angry. "You stupid student, have you learnt nothing of what I have taught you? When you bake plums in a pudding the ingredients change their identity. Due to the interaction of a weak cooking force, the plums change into cherries and pudding becomes a Black Forest gateau."

"Well how do you explain the plum that fell out of the pudding?" retorted the student; "There are no plums in a Black Forest gateau!"

"The gateau is unstable," barked the professor, "after a few minutes it falls apart and as it does so the cherries revert to plums."

How could the student argue? He was speaking to none other than the President of the Royal Society of Puddings. If he wanted to make it in his world and get a good career in the cake and pudding industry, he had to accept that plums baked in a pudding turn into cherries and the end result is a Black Fforest gateau...

Experimental physics suggests that a neutron is not a truly neutral particle. The evidence is that it does have charge but the effect of its charge has been cancelled by the coexistence of opposites. This occurs in atoms formed of charged particles. They do not display charge because they contain an equal number of opposite charges.

The neutron has magnetism. It has a magnetic moment of 1.91 nuclear magnetrons.[3] A particle cannot have a magnetic moment unless it has charge because the magnetic moment of a particle is created by the spin of its charge.

Then there is the issue of particle masses. If a neutron were a truly neutral particle it should be lighter, not heavier than a neutron. In *Quarks the Stuff of Matter* [4] Harald Fritzsch raised this issue when he wrote:

> "We do not understand why the neutron is heavier than the proton. Indeed, an unbiased physicist would have to assume the opposite by the following logic. It is reasonable to think that the difference in mass between the proton and neutron is related to electromagnetic interaction since the proton has an electric field and the neutron does not. If we rob the proton of its charge, we would expect the neutron and the proton to have the same mass. The proton is therefore logically expected to be heavier than the neutron by an amount corresponding to the energy needed to create the electric field around it."

Meanwhile, in the crazy world all was not going well for the mad professor of puddings. Another student baked plums in a pudding then licked it. He tasted a plum and so established that the plums did not become cherries when baked in a pudding. The mad professor was very upset by the arrival of that awkward fact. He ranted and raved, shouting that licking introduced too great a margin of error to be treated as a valid measure so the observation should be dismissed altogether.

The view of the neutron as a totally chargeless particle was challenged in 1957 by an experiment that revealed a very weak electric dipole on a neutron.[5] This was the equivalent in the story of the plum pudding being licked. On one spot the neutron displayed a minute negative charge – around a billion, trillion times weaker than that of an electron. This appeared to support the idea that the neutron is a bound state of opposite charges, which mostly cancel each other out rather than an entirely chargeless particle.

Physicists were quick to dismiss the discovery of an electric dipole on the neutron on the grounds that the measure of electric charge in the experiment was too weak to be taken as conclusive. In his textbook, *Nuclei and Particles*,[6] Emilio Segrèé pointed out that the electric dipole moment of the neutron, equal to the charge on an electron x10^{-20}, is so minute compared to the margin of error (i.e., 0.1 +/- 2.4) that "this moment could be exactly zero, in agreement with the theory".

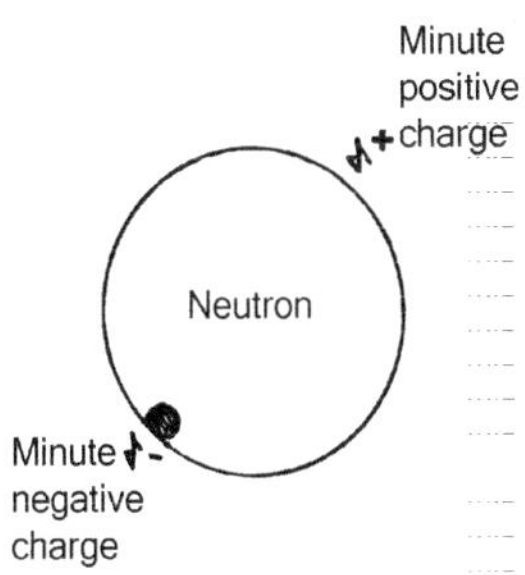

Fig 8 Electric dipole of a neutron

The electric dipole of a neutron would be expected to be very weak because if the neutron were an electron bound to a proton their opposite charges would all but cancel each other out but the measure should not have been treated as exactly zero. Though the margin of error indicated the measure was too weak to be conclusive it did not suggest there was no measure at all. The way Segrèé wrote about the result of that experiment was suggestive that he really wanted to dismiss the experiment because it revealed an awkward fact that needed to be discounted in order to protect 'the theory'.

Physicists argued against the bound state theory because the neutron displays the same value of quantum spin as a proton. That is to be expected. If an electron is locked onto a proton its quantum spin value would not be apparent. The quantum spin of the electron would be undetected, not lost, because an electron leaving a proton in beta-decay has the same quantum spin as an electron joining it in K-capture so angular momentum would be conserved in the overall process of formation and decay of neutrons.

Fig 9 Richard Feynman

The physicist and Nobel laureate Richard Feynman stated:

> "It doesn't matter how beautiful your theory is, it doesn't matter how smart you are. If it doesn't agree with experiment, it's wrong". [7]

Science is evidence based. In the history and philosophy of science it is unacceptable to disregard evidence in order to protect theories. When that occurs science is lost because its single most important rule is that theories should be ditched if discoveries or experiments prove they are wrong.

My father, Dr. Michael Ash, said to me:

> "Even if you find yourself standing alone against the entire world it does not mean you are wrong."

Fig 10 Michael Ash

Albert Einstein found himself standing alone against the entire world of physics when he challenged the uncertainty principle of Werner Heisenberg. Do you think he was wrong?

Do you think he deserved to be sidelined in physics because he rejected quantum mechanics? The scientific community thought he had lost his marbles.

Do you agree with them? Or do you think he was courageously upholding the integrity of science?

1. Matthews R., *Unraveling the Mind of God,* Virgin Books, 1992
2. Bernstein J., *Quantum Profiles,* Princetown University Press, 1991
3. Alvarez L. W; Bloch, F. *A quantitative determination of the neutron magnetic moment in absolute nuclear magnetrons.* Physical Review 57: 111–122. 1940.
4. Smith J. H., Purcell E. M., Ramsey N. F., *Experimental Limit to the Electric Dipole Moment of the Neutron,* Physical Review, 108: 120–122, 1957
5. Segrè E, *Nuclei & Particles* (1964) Benjamin Inc
6. Fritzsch H, *Quarks: The Stuff of Matter,* Allen Lane 1983.
7. Feynman R., Leighton R., Sands M., *The Feynman Lectures,* Addison-Wesley 1965

2. At the Still Point - Shane Guffogg

For me, this image invokes the still point of the atomic nuc-cleus around which electrons are orbiting - Ash

ESSAY 2

Quantum Craziness

"If you cannot - in the long run - tell everyone what you have been doing, your doing has been worthless."

- Erwin Schrödinger

Précis: Heisenberg and Bohr set a quantum bandwagon in motion. After sweeping aside Einstein, it took onboard most practicing physicists. This happened when a young Japanese theorist, Hideki Yukawa, used Heisenberg's formula to predict a particle that could carry a force to help keep the nucleus of the atom together. He called it a 'meson'. When mesons were knocked out of the nucleus of a silver atom in a cosmic-ray experiment, Heisenberg's uncertainty principle became universally accepted and the door slammed shut on Einstein.

Heisenberg's principle inferring uncertainty between time and energy allowed for particles to appear with a lot of energy if they existed for a very short space of time. Yukawa's particles were discovered to have massive energy and lifespans in the order of a quadrillionth of a second. These fitted Heisenberg's formula and so Nobel prizes were dished out, new theories abounded, and a search began to find more particles that could carry the forces of nature.

One who succeeded in the wake of Yukawa was Richard Feynman. Feynman was a brilliant young physicist from Caltech, the California Institute of Technology. Building on a theory from the nineteenth century, that light consists of waves of electric and magnetic force racing through space, he proposed that within the bounds of Heisenberg's uncertainty virtual particles of light could pop up alongside charged particles and exchange the force of electric charge, somewhat like broadsides between battleships. The suggestion was that energy could be borrowed from the Universe to bring the virtual particles into existence and so long as they vanished in time to repay the energy debt, within the time limit set by Heisenberg's formula, then accounts would be settled as though the Universe were a bank. When the question was raised that perhaps the force-carrying particles might get lost as the exchanges occurred over large distances, theorists suggested that they might feel or gauge for the presence of other particles in order to home in on their targets. These gauge theories were a big ask, but because particles were appearing and disappearing in high-energy experiments within the short spaces of time allocated by Heisenberg's formula it appeared to be sound, so they were accepted.

Essentially Heisenberg's principle was a licence for speculation. Physicists have speculated that the proton is capable of borrowing energy to create gluons for the binding of quarks, W and Z particles for the weak nuclear force, mesons for the strong nuclear force, virtual photons for electric charge and magnetism, gravitons for gravity and Higgs Bosons to account for mass. Attempts have been made to consolidate this borrowing from the universal energy bank into one grand unified 'GUT' account but it has not been successful.

The assumption that behind a shroud of uncertainty energy is borrowed and repaid to produce short-lived force-carrying particles is crazy, but because of the success of Heisenberg's formula it is accepted as the standard model in quantum physics. The credibility of physics hangs in the balance. If it became public knowledge that the uncertainty principle has been disproved, not only would the generally accepted theories for forces face collapse but confidence in quantum mechanics could evaporate. When quantum theorists ditched realism and common sense maybe they also lost sight of reason. Perhaps we should reconsider Einstein's stand for sanity and ask ourselves if his resistance to the uncertainty principle and his call for common sense and realism should be heeded?

~

Fig 11 Albert Einstein

Essay: According to Thomas Kuhn – widely recognised as one of the greatest authorities on the history and philosophy of science – Albert Einstein was the true father of quantum theory[1] and despite his deep friendship with Neils Bohr he was deeply unhappy at the direction quantum theory was taking under him. Bohr had come to dominate quantum physics as much through personality as physics. He had charisma and was very popular. Einstein was charming, humble and kind but he was always a loner whereas Bohr drew the younger men to him like a magnet. He was an unpretentious, energetic, much-loved man who was always concerned for the welfare of others. Through his powerful intellect and success as a fundraiser and administrator he established his Copenhagen Institute as a hub of theoretical physics. Nonetheless, many other people as well as Einstein have been deeply unhappy about Bohr's dominating influence over the burgeoning quantum theory. As Robert Matthews put it in Unravelling the Mind of God: [2]

> "The Copenhagen interpretation of quantum theory remains the most widely accepted way of looking at quantum theory amongst scientists today, but it should be stressed that, despite what some might claim, it remains just that: an interpretation. There is powerful experimental evidence that it is an acceptable way of looking at the world, but it is most definitely not the only possible interpretation."

Robert Matthews described Bohr's influence on quantum theory, that Werner Heisenberg dominated in the area of quantum mechanics, as a "quantum bandwagon". In his book he continued:

> "At the reins of the quantum bandwagon were one or two superstars - like Niels Bohr - and in the back were like minded theorists and experimentalists armed to the teeth with laboratory results to prove their case. If you get in the way of such a bandwagon, you will at best be left behind, and at worst be flattened."

Matthews quoted David Bohm as saying:

> "Most physicists said they followed Bohr without really knowing what Bohr was doing."

1925 to 1928 were the years of Knabenphysik, 'the boy physics'. Radical new ideas for quantum theory were emerging from the minds of Werner Heisenberg, Wolfgang Pauli and Paul Dirac, amongst others. Pauli was famous for his exclusion principle and Einstein was excluded. The older generation, including Bohr, Einstein and Schrödinger, struggled to keep up with the bright boys. Bohr was revered by the younger men. He enjoyed their friendship and was indefatigable in clarifying the detail of their emerging theory of quantum mechanics. It is no wonder the new generation of physicists put more store by Bohr than by Einstein.

Sadly the greatest scientific genius in history was swept aside in the storm of theoretical physics pouring out of young tempestuous heads. In a spoof on

Faustus, performed at the Copenhagen Institute of Theoretical Physics in 1932, Paul Dirac ended the play with: "Old age is a cold fever that every physicist suffers with; when he is past thirty he is as good as dead." [3] In the eyes of the emerging physicists Einstein was a has-been. He refused to acknowledge that at its root the world was essentially unknowable. Einstein pioneered quantum theory and led it for many years then became its most redoubtable critic. It is easy these days to overlook that quantum theory originated from Planck and Einstein not Heisenberg and Bohr. People tend to forget Einstein was the leader of the quantum revolution, the brightest star in the original quantum firmament.

The quantum world began to lose touch with normality even under Einstein's watch when waves were appearing as particles and particles were behaving as waves. People were finding it increasingly difficult to comprehend quantum reality. When Bohr took over, realistic models were rejected as naïve, and unrealistic models were emerging as fast as the uncertainty formula allowed. These models were mathematical constructs that worked amazingly well but in the real world they made no sense at all. As Heinz Pagels said: "Something inside us doesn't want to understand quantum reality. Intellectually we accept it because it is mathematically consistent and agrees brilliantly with experiment. And yet the mind is not able to rest."[4] Craziness was becoming the quantum norm, so much so that Niels Bohr said, at the end of a lecture given by Wolfgang Pauli in 1958: "We are all agreed that your theory is crazy. The question which divides us is whether it is crazy enough." [5]

In the nineteenth century, light was described in terms of perpendicular electric and magnetic waves and in the twentieth century the electromagnetic model was applied to the photon particle of light, which could be imagined as perpendicular wave trains, one carrying the electric force and the other the magnetic force. The idea that forces could be carried by particles was a natural consequence of this thinking. Heisenberg's principle inferred that within the bounds of uncertainty the energy available for a force carrying particle could be more if the time over which it was effective was less; in other words, the shorter the life span of a force-carrying particle, the greater the force it could carry. This is where quantum mechanics began to show its teeth.

Fig 12 The Photon

Fig 13 Hideki Yukawa

When Rutherford discovered the atomic nucleus was populated by protons that strongly repelled, physicists realised there must be a very powerful force bonding them together. Nobody knew what that could be until 1934, when a young Japanese theorist, Hideki Yukawa, predicted the existence of a force-carrying particle exchanged between protons, which he dubbed a meson. Using Heisenberg's principle of uncertainty, he calculated how short lived it would need to be in order to have sufficient energy to overcome the repulsion between the protons. It would have to come into being in one proton and undergo an exchange with another in less than a trillionth of a second before decaying. The suggestion was that energy could be borrowed from the Universe, and used to enact a very powerful force of binding between protons in the atomic nucleus over an incredibly short space of time, before the energy debt was repaid to the Universe by the demise of the force-carrying particle. The idea was crazy but with realism dismissed, that didn't matter. Beneath the shroud of uncertainty it could happen because the uncertainty principle ensured no one could prove otherwise and the mathematics around it was brilliant.

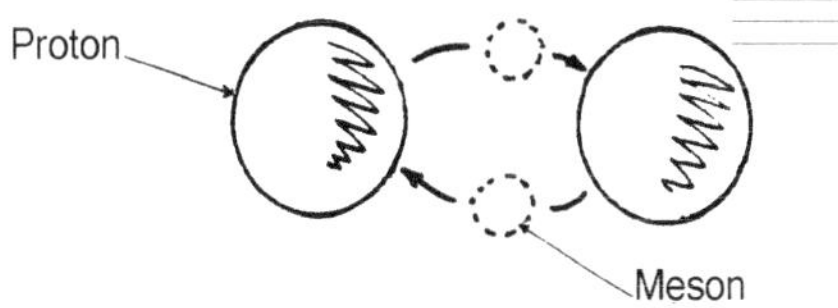

Fig 14 Mesons binding protons

Young physicists were having fun. They didn't want the old folk to come in and spoil their party. You can begin to see why, when the Nobel Prize was presented to Chadwick in 1935 for the discovery of the neutron, the new generation in physics didn't want Rutherford's account for it. They turned their back on him, and on Einstein, and went hunting for mesons instead.

Fig 15 Cecil Powell

In 1938, a young research assistant and lecturer at Bristol University, called Cecil Powell, began to litter the tops of Welsh mountains with photographic plates in black paper wrapping. He then sent them high in the sky attached to weather balloons. He was looking for something very special and very rare. He wanted to catch a high-energy cosmic-ray particle from the Sun colliding with the nucleus of a silver atom in the photographic emulsion. He had to send his photographic plates up into the rarified atmosphere of high altitudes because that is where cosmic rays were most plentiful. They could penetrate the paper wrapping while light could not. His research technique was amazingly simple and ingenious but there was one snag. Because atoms are mainly space he would have to send a lot of photographic plates into the heavens, and it would take an army of girls and assistants working in the attic of the W.D. & H.O. Wills building in Bristol to develop and analyse them. He persisted for nine years before, in 1947, he got the snap of a lifetime that would win him the Melville Wills Professorship of Physics at Bristol University in 1948, Fellowship of the Royal Society in 1949, and the Nobel Prize in 1950. He caught mesons on camera.

A young girl called over a postgrad to check what she had spotted on a photographic plate. He was visibly excited as he looked closely at the tracks in the photographic emulsion. He ran off to collect Cecil Powell and on close examination they concluded they had hit the bull's eye. The photograph witnessed an alpha-particle (helium nucleus) with three thousand billion electron volts of energy slamming into the nucleus of a silver atom. Out of the wreckage of the nucleus a shower of 140 particles had appeared, which fitted Yukawa's predicted mesons. From the lengths of the tracks they ascertained that the new particles – that came to be known as 'pi-mesons' or 'pions' – survived for only a quadrillionth of a second (10^{-15} Sec.) This concurred with the time constraints inferred in Heisenberg's formula. Hideki Yukawa was awarded the 1949 Nobel Prize for physics, and quantum mechanics came of age.

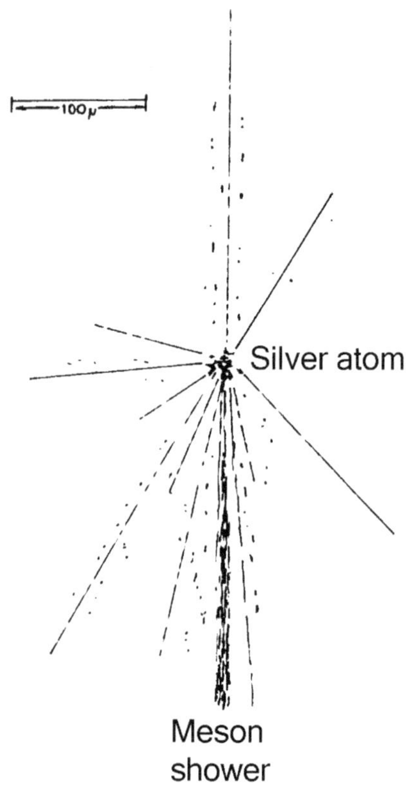

Fig 16 The cosmic ray photograph

Fig 17 Richard Feynman

In high-energy experiments, particles were appearing and disappearing within incredibly short spaces of time. These fitted the time allowance in Heisenberg's formula for what was allowed in terms of force-carrying particles to come into existence and disappear again within the constraints of uncertainty. This enabled Richard Feynman, a brilliant young physicist from the California Institute of Technology, to propose that within the bounds of Heisenberg's uncertainty virtual particles of light could pop up alongside charged particles and exchange the force of electric charge between them, somewhat like broadsides between battleships. After the exchanges with other charged particles, as the virtual photons vanished within the time limits of Heisenberg's uncertainty formula, they could exist but they could never be detected. Feynman was clear that they were more mathematical constructs than real entities. This is why, in the system known as quantum electrodynamics (QED), they were called 'virtual photons'.

Richard Feynman drew diagrams to illustrate the exchange of virtual photons. These Feynman Diagrams have also been used to illustrate the exchange of other force-carrying particles to explain the action of other forces in the theories of quantum mechanics. However, the concept of force-carrying particles being exchanged over large distances met with certain difficulties. Once a force-carrying particle launched into space how could it locate its target particle at a distance? This was essential if it was to cause an interaction, decay and quickly repay the debt of energy incurred in its formation within the time constraints of the uncertainty principle. Energy conservation laws didn't allow for a single force-carrying particle ever to be lost.

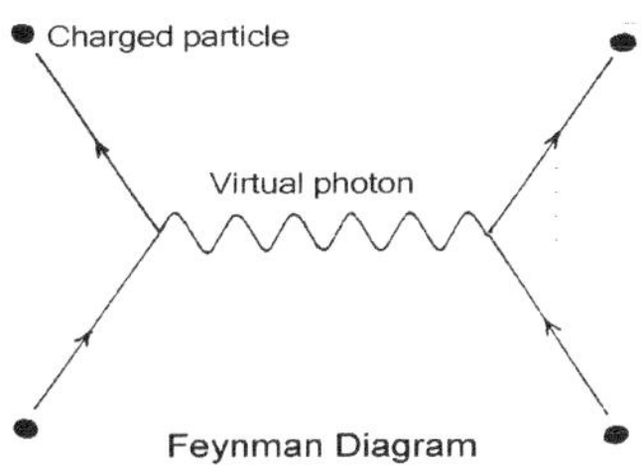

Fig 18 Feynman diagram

To overcome the problem, gauge theory was introduced, which suggested that force-carrying particles could gauge the force; that is they could feel for the presence of another force-carrying particle and so home in on their

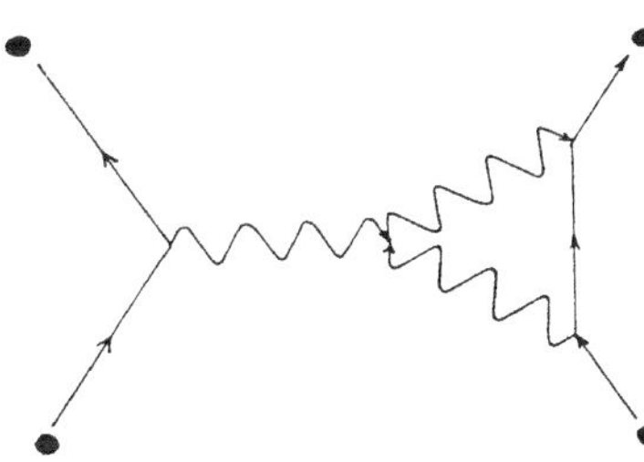

Fig 19 Gauge theory diagram

target. There are problems associated with this theory. The idea of particles homing in on each other over short ranges could be imagined but using gauge theory to account for electromagnetic fields operating over the infinite distances of outer space beggars belief. It is hard enough to imagine the instantaneous exchange of virtual photons through deep space, let alone their ability to gauge for each other over vast distances. Quantum mechanics may satisfy the math but it defies common sense.

The principle of uncertainty gave theorists licence to speculate. The idea, that subatomic particles run credit with the Universe, as though it were a bank, is crazy. In the Talmud it is written, 'We see the world, not the way it is, but the way we are'. With their mortgage, bank loan and credit cards, professional physicists appear to have projected the image of credit on the subatomic world in which they imagine Mr Proton borrowing energy to create gluons for the binding of quarks, W and Z particles for the weak nuclear force, mesons for the strong nuclear force, virtual photons for the forces of electric charge and magnetism, gravitons for the force of gravity and Higgs bosons to account for mass. Attempts have been made to consolidate the borrowing from the universal energy bank into one grand unified 'GUT' account. This has not been successful.

Fig 20 Mr Proton

The investment in Heisenberg's uncertainty principle is enormous. If the neutron reveals it to be invalid, then the entire edifice of quantum mechanics would collapse like the proverbial house built upon sand and that is unthinkable to the modern physicist. In the words of Anthony J. Leggett:

> "Quantum mechanics...has had a success which is almost impossible to exaggerate. It is the basis of just about everything we claim to understand in atomic and sub-atomic physics, most things in condensed-matter physics, and to an increasing extent much of cosmology. For the majority of practicing physicists today it is the correct description of nature, and they find it difficult to conceive that any current or future problem of physics will be solved in other than quantum mechanical terms. Yet despite all the successes, there is a persistent and, to their colleagues, sometimes irritating minority who feel that as a complete theory of the Universe, quantum mechanics has feet of clay, indeed 'carries within it the seeds of its own destruction'. "[6]

The credibility of science is at stake. The feet of clay in quantum mechanics is Heisenberg's theory. For nearly a century the world of physics has endorsed it, even after Albert Einstein warned it was wrong. Physicists dismissed Einstein because of his warning and discounting the evidence against the theory that should have vindicated him. They then used the uncertain theory as the foundation of quantum mechanics. These are the seeds that could destroy quantum mechanics. The scientific method has been debased and the scientific criteria for truth undermined. Critics could argue that scientists have been bent on proving what they believe rather than the unbiased pursuit of truth, which is the criterion of a religion not science. Many people believe we create our own reality. They would have grounds to argue that experiments merely reflect the beliefs of the scientists performing them. It is no wonder any mention of the neutron being a bound state of electron and proton upsets physicists. If the neutron were an electron immobilised by a proton, then the outstanding success of quantum mechanics will have revealed just how effective science can be in substantiating stories.

For all its successes quantum mechanics could be likened to a car that has failed its test of roadworthiness. It may run really well but on even a single item of failure it has to be repaired or scrapped. The problem can't be ignored. This rule applies to quantum mechanics as well as car mechanics.

The Austrian philosopher, Karl Popper, used an analogy of black and white swans to explain that science is not in the business of proving theories but rather of disproving them. Someone could have a theory that all swans are white but even if a thousand white swans were counted, their belief would not be proved and the addition of more white swans wouldn't make it any truer. However, the appearance of a single black swan would disprove the theory altogether. Scientific theories are white swan belief systems, which survive until they are destroyed by the appearance of a black swan, that is, the arrival of even a single fact, which makes it clear that they cannot be true.[7] The integrity of science depends upon that but when the neutron appeared as a black swan over the quaint little pond of quantum mechanics, instead of welcoming its arrival – as scientific integrity would demand –- physicists tried to shoot it. Had they allowed it to land, Heisenberg's principle would have been a dead duck and the virtual force-carrying particles, their darling little ducklings, would have certainly drowned. Failing to submerge it, they muddied the waters with incomprehensible math and whitewashing it with quark they made a flap about quantum spin and chased it away.

It is the story of the Emperor's New Clothes all over again. The virtual particles of quantum mechanics are phantoms but people in physics won't admit to it because there are too many jobs and academic reputations at stake. However, there is no future for science if awkward facts are swept under the carpet to save the theories. In the words of George Gamow:

> "Staggering contradictions of this kind, between theoretical expectations on the one side and observational facts or even common sense on the other are the main factors in the development of science." [8]

Fig 21 Stephen Hawking

At his inauguration as Lucasian professor of mathematics at Cambridge University, Stephen Hawking gave a lecture entitled, *The End is in Sight for Theoretical Physics.* In it he said that because of the Heisenberg uncertainty principle, the electron could not be at rest in the nucleus of an atom.[9] The fact is that experimental physics reveals electrons are at rest in the nucleus of an atom and, because of that, the end is in sight for theoretical physics.The neutron is the particle used by physicists to detonate weapons of mass destruction. It could be seen as divine retribution if the neutron were to blow physics apart.

There is a silver lining to the Heisenberg cloud. If through the uncertainty principle he has exposed uncertainty in the scientific criteria for truth, he will have released humanity from the burden of truth imposed by science. Religions have a history of telling people what they can and cannot believe. Increasingly, there is a tendency for scientists to impose their understanding of the Universe on the lay public which is engendering an attitude of intolerance unbecoming of science. If the neutron unhinges the Heisenberg uncertainty principle scientists will no longer command the high ground in regard to the truth. As waves breaking endlessly on the shore so future generations will be able to throw interpretations of truth on the beach of uncertainty. No longer will anyone in science or religion be in a position to say for certain, 'I am right and you are wrong'. Heisenberg's lasting legacy could be his contribution to the infinity of possibility. He will have helped in the emancipation of humanity from the burden of scientific as well as religious truth, so that every human spirit can be at liberty to believe whatever he or she chooses. Through the philosophy of uncertainty Heisenberg may have done more for the truth than any saint or scientist because the paradox is the less certain we are of truth the closer we are to it.

1. Kuhn T., *Black-Body Theory and the Quantum Discontinuity*, Clarendon Press, 1978
2. Matthews R., *Unraveling the Mind of God* Virgin, 1992
3. Segrè G., *Faust in Copenhagen: A Struggle for the Soul of Physics*, Viking, 2007
4. Pagels H., *The Quantum Code*, Michael Joseph, 1982
5. Ratcliffe S. (Ed) *Oxford Essential Quotes 2014*
6. Leggett A.J., *The Problems of Physics*, Oxford University Press, 1987
7. Popper K, *The Logic of Scientific Discovery* Hutchinson, 1968.
8. Gamow G., *Thirty Years that Shook Physics*, Heinemann
9. Hawking S., *Blackholes and Baby Universes*, Bantam, 1993

3. As Above So Below - Shane Guffogg

Here I see images of energy interlacing with the cosmos - Ash

ESSAY 3

Quark Nonsense

"Will some unknown young scientist find a new way of looking at fundamental physics that clarifies the picture and makes today's questions obsolete."

Murray Gell Mann

Précis: Cosmic-ray research was haphazard. It depended on a cosmic-ray particle with just the right energy slamming into the nucleus of an atom. This sort of bullseye was very rare because atoms are mainly space, so in the aftermath of Yukawa and Powell's joint success – for which they both received Nobel Prizes – physicists and engineers designed and built colossal machines that could accelerate particles with controlled energies and slam them with precision on designated targets. Quickly they began to churn out a veritable zoo of particles, which were mostly very short lived. To make sense of them, theorists hypothesised that they were composed of triads of smaller particles, which they called 'quarks'.

Quark theory is full of speculations. For one thing quarks are assumed to have fractional charge when fractional charges have never been observed in nature. Then there is the problem that quarks have never been detected in a free state, not to mention that they have been calculated to be up to five times more massive than the particle they composed. Two minuses make a plus. Physicists used these two problems to cancel each other out. They knew that after nuclear fission as protons became more tightly bound together they lost a small amount of mass, so they postulated that quarks had lost fourteen-fifteenths of their mass when they became bound in a proton and so were so strongly held together it would be impossible ever to isolate one. Then there was the problem of particle longevity. Most of the particles synthesised in high-energy labs, that were supposedly made of quarks, lasted for quadrillionths or trillionths of a second, billionth's at the most, whereas protons have an estimated longevity of quadrillions times quadrillions of years. Physicists have been unable to explain this considerable discrepancy in lifespans. Quark theory was living up to its name. In Germany quark is slang for nonsense.

Quark theory protects the uncertainty principle. It has enabled physicists to explain away the neutron as a chargeless particle. By claiming that when an electron is attracted to a proton an up-quark transmutes into a down-quark and the electron and proton disappear to be replaced by a chargeless neutron, they managed to overcome the nagging problem of discounting the obvious structure of a neutron. Without quarks they would have to accept it is an electron locked onto a proton and ditch uncertainty as the foundation of quantum mechanics. If they did that, everything they believe about particles and forces would be lost. Quantum physics would have to be rewritten. Quantum mechanics would suffer an existential hit.

Physicists claimed they discovered quarks in protons after bombarding them with high-energy electrons. But they didn't actually detect quarks in the protons. What happened was some of the electrons ricocheted as though they struck something hard inside the protons. Quarks were just an interpretation of these events. In cosmic-ray and high-energy research protons have been shown to contain mesons and mesons could have been responsible for the ricochets. Science is on very shaky ground. There is a real need for a clean sweep in particle physics and a new underpin for quantum mechanics. A new theory needs to be found to replace the uncertainty principle if physics is to be saved from a wipeout.

~

Essay: In the wake of Cecil Powell's spectacular success in cosmic-ray research, scientists and technicians designed and built machines with which it was possible to accelerate particles to high energies to bombard atomic nuclei. Cosmic rays were unpredictable and haphazard in their energies whereas, in the particle accelerators, physicists could control the energies of their bullet particles and target them on atomic nuclei with precision. This was the advent of high-energy research and the particle accelerators came to be known as high-energy laboratories. The outcome of this research combined with quantum mechanics came to be known as the New Physics. The crowning achievement of high-energy nuclear research was an entirely new understanding of matter described as quark theory.

On 27 January 1977, a BBC television programme, *The Key to the Universe*, reported on the New Physics. In a companion book to this programme, under the same title, Nigel Calder wrote about how high-energy research led to the theory for quarks:

> "In the early 1930s the contents of the Universe seemed simple. From just three kinds of particles, electrons, protons and neutrons, you could make every material object known at the time. Thirty years later human beings were confronted with a bewildering jumble of dozens of heavy, apparently elementary particles, mostly very short lived. They came to light either in the cosmic rays or in experiments with the accelerators. The particles had various mass-energies and differing qualities such as electric charge, spin, lifetimes and so forth. Moreover they were given confusing, mostly Greek, names so that one of the most eminent of physicists, Enrico Fermi, was driven to remark before his death in 1954, "If I could remember the names of all these particles I would have been a botanist".
>
> "The proliferation could be understood, to some extent, in that many of the particles seemed to be energetic relatives of the proton. Because they possessed greater inherent energy their masses were greater. Each was in some sense less tightly bound together than a proton and it could quickly change into a proton with a release of binding energy and an associated loss of mass. But that implied the proton was not a truly basic particle: it was made of something else, which could be bound more or less tightly together."

Fig 22 Murray Gell Mann

"A small group of theorists brought order out of chaos. The principle figure amongst them was Murray Gell-Mann of Caltech (The California Institute of Technology), then in his early thirties. He declared that all the heavy particles of nature were made out of three kinds of quarks. He had the word from a phrase of James Joyce 'Three quarks for Muster Mark.' It was the mocking cry of gulls, which Gell-Mann took as referring to quarts of beer, so he pronounced quark to rhyme with 'stork'. Many other physicists rhymed it with 'Muster Mark'. In German, as sceptics were not slow to notice, 'quark' rhymed with Mark was slang for nonsense...." [1]

There is a lot in a name and the quark theory is nonsense, as the name suggests, not because of mocking gulls but because of a black swan. Karl Popper's analogy of black and white swans, used to explain that science is not in the business of proving theories but rather of disproving them,[2] can be applied to quark theory. Quark theory is a white swan belief system that is doomed because there is a black swan, a single obvious fact, which makes it clear that the theory cannot be true.

The black swan for the quark theory is the proton. The single fact, which throws the quark theory into question, is the disparity in life span of the proton and practically all other particles discovered in the twentieth century. The lifespan of a proton has been determined at ten billion, trillion, trillion years, whereas one ten-billionth of a second is considered a strangely long lifespan for any of the new particles found in high-energy research. As spontaneous proton decays have never been observed, the proton can be treated as being infinitely more stable than any of the new, heavy particles that have come to light in the accelerators. Physicists call the heavy particles 'baryons' from the Greek word for heavy. They account for the difference in lifespan between protons and other baryons in a law called the law of conservation of baryon numbers. However, this law doesn't tell us why protons live for so much longer than the other particles it simply states, mathematically, that they do so. It is just one of the many instances in science where 'This is how it is' is stated without saying, 'This is why it is'. Mathematics in physics can account for how things work but there is a general lack in physics of insights for why things are the way they are.

With over 31 million seconds in a year, the difference between the estimated lifespan of a proton and one of the new particles is a figure in excess of 3×10^{50} (3 with fifty zeros behind it). Mathematicians use the number 10^{50} as a cut off point for probability. If the probability of something happening is less than 1 in 10^{50} then scientists are forced to accept that it never occurs. With a difference in stability in excess of 10^{50} it can be argued that is is improbable that protons decay so it is beyond the bounds of reason to treat them the same as other baryons, unless good reason is given for the disparity in their life spans.

This point is borne out by a building site analogy. Imagine walking down a road between two building sites. On the site to the left, the houses disintegrate as soon as they are built. Within a fraction of a second of the last lick of paint being applied, they vanish. On the site to the right the houses are advertised for sale with a trillion-year guarantee. It would be nonsensical to assume that identical bricks and mortar and construction techniques are employed on both building sites without providing an account for the difference in the durability of the houses. Yet, in physics today, physicists claim protons and all other baryons are constructed of quarks cemented together by gluon bonds.

There are believed to be six different types of quark bound in triads by gluon

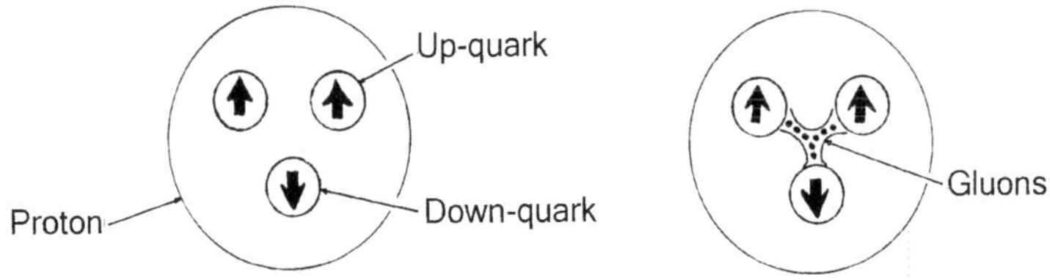

Fig 23 Quarks

bonds. People who believe the quark theory accept that different combinations of three quark bricks, cemented together with gluon bonds, give rise to all the baryon houses, including protons. But they don't explain why protons are infinitely more durable than all the other particles in the particle zoo. Quarks could be used to explain protons, or they could be used to account for the other baryons, but not both. To use quark theory to explain only protons would be pointless, because the theory was invented to explain the particle zoo. But if quark theory provided an explanation for all the baryons apart from the proton it would still be pointless because the great bulk of the mass of the physical Universe consists of protons, whereas the other baryons – apart from the neutron – have been observed only in high-energy experiments. If the theory were to exclude protons the entire worldwide programme of high-energy research, which dominates physics, would be meaningless.

Physicists talk about discovering new particles of matter in the high-energy research but to say they have been discovered is misleading because discovery implies that something already exists and is waiting to be found. The short-lived baryons have been created out of the massive amounts of energy fed into the high-energy particle accelerators. These new heavy particles have been synthesised in an accelerator. They have no place in normal matter. They are just anomalies of high-energy research.

When the discovery of the top-quark was announced at Fermilab in April 1994, Jim Dawson opened his report in the Minneapolis Star Tribune with the following statement:

"So there we have it, after more than two thousand years of searching, all of the fundamental stuff of Democritus' atom has been revealed. The crowning moment came a couple of weeks ago when physicists announced that a gigantic, 5,000-ton machine apparently had detected a very small particle called the top-quark." [3]

The physicists at Fermilab in Illinois didn't actually see a quark. What they

Fig 24 Fermilab

saw was the tracks of a jet of electrons and muons that they supposed to be the breakdown products of W-particles, which they took to be remnants of a top-quark. The 'discovery' of Higgs boson at CERN in 2012 was an interpretation of similar fallout. No one actually saw the boson because it decays too quickly to be detected. What they saw was debris from high-energy proton collisions which they interpreted as a strong probability of a Higgs boson decay. Quarks are supposed to be stable but while everything in the world is supposed to be made of quarks, not a single quark has ever been observed in a free state – even though no expense or effort has been spared in the effort to find one.

Physicists reason that a quark has never been observed in the free state is because it is impossible to have a gluon bond without a quark attached to either end.[1] They speculate that if a quark is blown out of a proton in a high-energy impact, two new quarks form out of the energy of impact. A new quark forms at the broken end of the gluon bond attached to the proton to restore the proton to three quarks, and a new quark forms at the end of the split gluon bond attached to the quark knocked out of the proton to form a two-quark meson. This is how physicists account for the mesons that appeared in Cecil Powell's cosmic-ray experiment. This account is reminiscent of the mediaeval argument that if an angel were cut in half, because it is impossible have half an angel two new halves would immediately grow on either end of the split angel; so the outcome would be two entire angels.

Another problem with the quark theory is each quark in a proton has been estimated to have five times the mass of the proton itself. Physicists got round that problem by suggesting that when quarks form a proton they lose fourteen-fifteenths of their mass. As their equations show nuclear binding increases with a minute loss of mass, physicists contend that the quarks would then be so tightly bound that a quark in the free state could never be detected. That, they say, is another reason why no one has ever seen a quark.[1] Many physicists are unhappy about the discrepancy between the mass of the quark and the mass of the proton. Richard Feynman is on record saying, "The problem of particle masses has been swept in the corner." [1]

In the late 1960s and early 70's a series of experiments were performed at the Stanford Linear Accelerator in California (SLAC) in which electrons were

Fig 25 SLAC

accelerated down a three-kilometre long 'vacuum tube' by intense radio pulses and then targeted on protons in liquid hydrogen. The results showed electrons being scattered, or bounced back, from what appeared to be something small and hard in the protons of the hydrogen atoms. From this it was inferred that the protons contained smaller particles.[1] Physicists were looking for quarks and naturally concluded that their bombarding particles were bouncing off quarks in the proton so these experiments caused the quark theory to become accepted into mainstream physics and led to a Nobel Prize for Gell-Mann in 1969. The SALC experiments appeared to be supported at CERN, the European Nuclear Laboratory near Geneva. At CERN, protons were being accelerated in vast intersecting rings.

At the intersections, energetic protons were directed into head-on collisions.

Fig 26 CERN

In the high-energy impacts new particles were formed and came out at right angles to the beams. Again these results seemed to confirm the quark model. However, the SLAC and CERN experiments did not necessarily prove the existence of quarks within the proton. They merely demonstrated that the proton has a 'hard heart'. The hard heart of a proton could result from its containing not quarks but mesons. Whereas quarks have never been knocked out of protons, in his classic cosmic-ray research Cecil Powell did knock mesons out of protons.

The hard scattering observed at SLAC and CERN could have been caused by impacts with mesons not quarks.

During World War II physicists succeeded in harnessing the energy locked in the atom to make the most terrible weapons of destruction ever known to mankind. Governments sprinting in the arms race endowed them with virtually unlimited budgets for continued nuclear research. Vast sums of money have been spent on building, maintaining and running massive particle accelerators. Using these, scientists have created a host of new particles accounted for in a quark theory. The theory threw up new difficulties and exciting predictions, which required more research and more powerful accelerators. The

Large Hadron Collider at CERN in Geneva, costing over eight billion euros, is the biggest and most powerful accelerator in the world. This monster machine is twenty times more powerful than any built previously. Its main purpose was to identify the source of mass – believed to be the Higgs boson - and to simulate the first moments of the Big Bang. It was also built to further research into quarks and force-carrying particles. Creating a new generation of particles this accelerator will lead to more elaborations in the theories, which will require more research and eventually the need for an even bigger accelerator. To prove the predictions of the Grand Unified Theory physicists would need an accelerator as big as the solar system.

High-energy research has created an endless cycle in which research physicists create problems for theoretical physicists to solve, and they, in turn, predict yet more problems to research. While this ensures employment for scientists and their army of technicians, it produces information of little relevance or value to the tax payers who have to foot the enormous bill. It is time for a change in physics. Appearing as pillars of truth and harbingers of change, the evidence provided by the neutron and proton could bring quantum mechanics into line with reason and common sense, more in step with the thinking of Einstein than Bohr and Heisenberg.

If experimental physics is to be believed, neutrons are formed of electrons attached to protons. If that were so, Heisenberg's uncertainty formula would predict that the certainty of their positions would lead to a wide range of electron velocities. Some would be creeping while others would be racing at nearly the speed of light and most would have a velocity somewhere in between. That would be reflected in a wide range of neutron masses. Experimental physics tells this is not the case. All neutrons have precisely the same mass. That suggests Heisenberg's principle is wrong.

Fig 27 Fermilab

Science is supposed to be evidence based but, for the best part of a century, physicists have discounted the evidence of the neutron in order to save the uncertainty principle and they appear to have invented quark theory to establish that neutrons are entirely chargeless particles. They built basilicas with rings in front – reminiscent of the Vatican – to prove it, and from their vaunted position they preached that a weak force changed up-quarks into down-quarks and that the Universe depends for its existence on their observing it. When Einstein warned that Heisenberg's uncertainty formula was a witches calculus and the Universe has objective reality and doesn't depend on human observation for its existence, the priesthood in the New Physics denounced his call for realism as being an obsolete prejudice.

Do you think the New Physics is in full alignment with the ethics and philosophy of science?[4] Do you trust that high-energy research has actually proved the existence of quarks, Higgs bosons and the virtual particles of quantum mechanics, or are they just blinding us with science? Do you entertain the possibility that physicists don't believe in objective reality because they are endeavouring to prove what they believe rather than what is objective reality? For the latter part of his life Einstein was a lone voice in the wilderness crying for common sense and realism in quantum physics. Do you think it is possible he may have been right?

1. Calder N., *Key to the Universe: A Report on the New Physics* BBC Publications 1977
2. Popper K., *The Logic of Scientific Discovery* Hutchinson 1968
3. Dawson J., *Star Tribune of Minneapolis*, May 15th 1994
4. Solovine Maurice 1987 *Albert Einstein Letters to Solovine*, New York Philosophical Library.

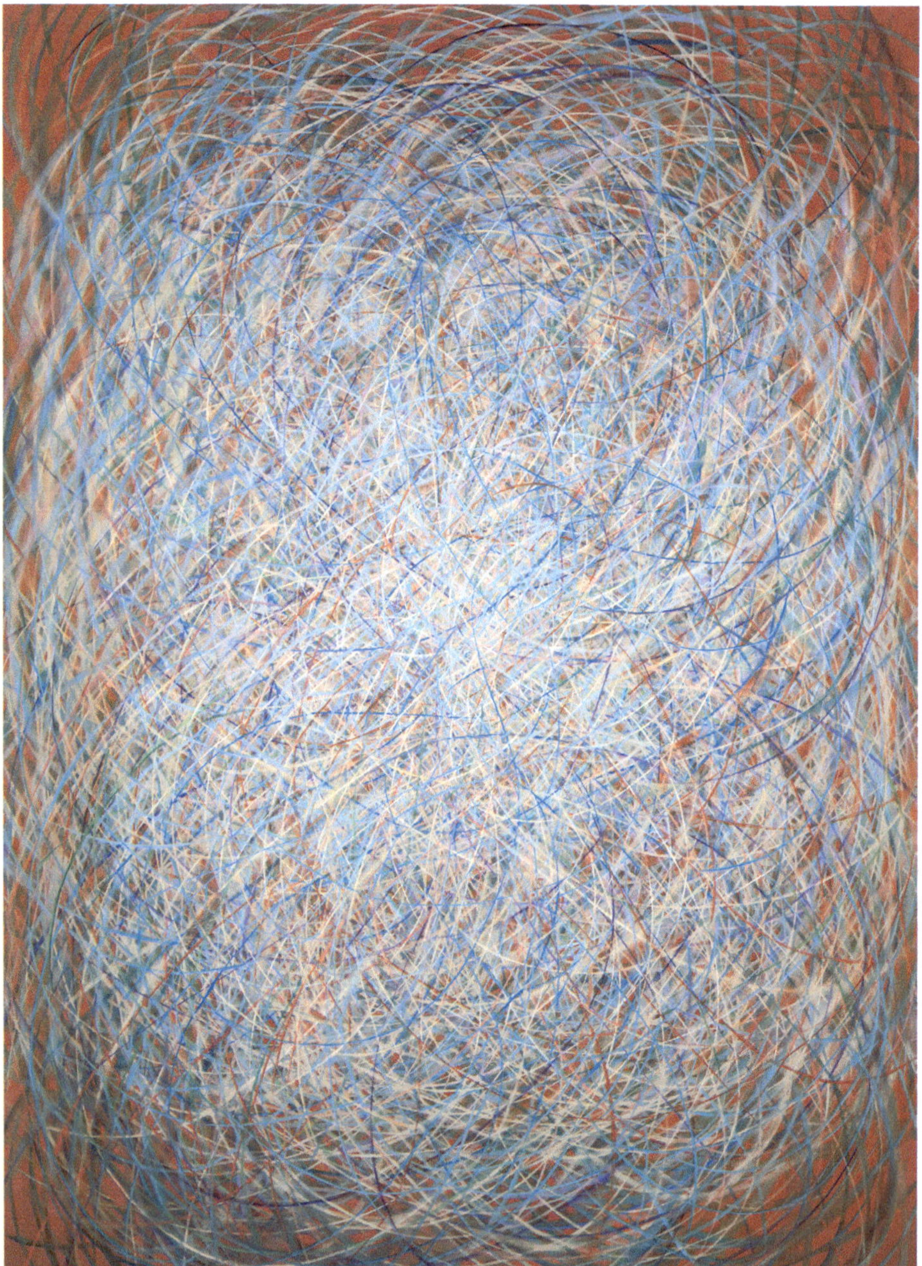

4. Reach into the Silence - Shane Guffogg

This image is quite the best artform I have seen suggesting the silent spin of vortex energy - Ash

ESSAY 4

The Vortex of Energy

"I believe that matter itself is just spin."

Eric Laithwaite

Précis: Einstein spent the second half of his life in an unsuccessful search for a unified field. He might have succeeded had he considered a discarded theory from the nineteenth century. In Victorian Britain there was a theory that the atom was a vortex in the ether. It explained many of the properties of matter but had been abandoned because it couldn't account for new discoveries about the atom. Neils Bohr succeeded where the vortex atom theory had failed and his success catapulted him into a position where he took the lead from Einstein in quantum physics.

I came across the vortex theory in 1965, not from reading physics but from reading a book on yogic philosophy. The book detailed the theory of the vortex atom and how it fitted yogic philosophy because in ancient India yogis had declared that the smallest particles of matter were vortices of prana. I was only sixteen at the time but I knew enough about physics to realise the enormity of what I had discovered. The book I borrowed from an antiquarian library came out in January 1905, several months before Einstein published his famous equation $E=mc^2$. I was staring at print on a page that prioritised Einstein because I knew prana was the Indian word for energy. If long before Einstein mystics in India knew that matter was a form of energy, maybe their description of the form was correct. My mind was racing. In the nineteenth century physicists thought the atom was the smallest particle of matter. It was only in the twentieth century that subatomic particles were discovered. I realised subatomic particles could be vortices of energy so when I entered University I decided to test that idea against modern physics.

When I applied the vortex theory to subatomic particles it worked immediately. The spin of energy could explain the tendency of mass to stay put. Gyroscopes and spinning pebbles skipping on pond surfaces demonstrate the way spin sets up resistance to movement out of the plane of spin. In a ball vortex, modelled by a ball of string or wool, spin is on infinite planes. If a corpuscular subatomic particle was a ball vortex of energy it would resist movement in any direction. That could set up the static inertia characteristic of mass. It could also account for the three dimensional extension of matter and for positive and negative charges. They could be caused by the spin of energy either in or out of the ball vortex. It struck me electric charge could be an expression of the dynamic state of the vortex. If charge was the action of a unitary vortex of energy it could not be fragmented. Every vortex particle would be a whole unit of charge. I believed that because in my college physics I was taught there are no partial charges. I learnt that protons have a positive charge and electrons have a negative charge and opposite charges cancel each other out so it was obvious that if an electron and proton came together to form a neutron it would have no apparent charge. For my theory to hold the neutron had to be a proton vortex and an electron vortex that had come together and lost their apparent effect by the coexistence of two opposites. Nonetheless, my physics lecturer insisted that a neutron was not an electron docked on a proton. He did not say why. He simply said that was the way it was and I had to believe it if I wanted to pass my exams. That was when I decided to read the subject for myself, free of confusing theories, unfounded prejudices and bewilderingly complex mathematics.

In subsequent years I realised the vortex theory explained away materialism and upheld Einstein's world view that the Universe has no material substance. It contributed to the stream of knowledge heading toward a non-mechanical view of reality; the view that the Universe is more like a mind than a machine.

Fig 28 Albert Einstein

~

Essay: In his later years Einstein worked tirelessly on a field theory to unify gravity and electromagnetism but his attempts were unsuccessful. He was becoming increasingly detached from new experimental discoveries and chose to ignore other forces. In the words of Walter Isaacson from *Einstein: His Life and Universe*[1]:

"... in his quest for a unified theory, there seemed to be a lot of mathematical formalism but few fundamental physical insights guiding him."

I believe Einstein could have succeeded if he had adopted a fundamental insight that was available and could have guided him. His equation $E=mc^2$ provided the mathematics that equated energy and mass but he lacked a visualisation of how energy forms mass. A model that had the potential to achieve that was central to physics in Britain during the latter half of the nineteenth century and was taught at Cambridge until 1910. It is a pity he didn't consider it because it had the power to unify gravity and electromagnetism. It needed to be revised and updated because it had been applied to the wrong thing and at the wrong level but in principle it was an extraordinary insight that should never have been lost.

Fig 29 Herman von Helmholtz

In the nineteenth Century the German physician and physicist, Herman von Helmholtz, who coined the term 'quantum', proposed that atoms were vortices in the aether (ether). His historic theory of the atom as a vortex was championed by Lord Kelvin (William Thomson) who was a leading figure in physics toward the end of the nineteenth century. In Kelvin's day, atoms, thought to be the smallest particles of matter, were commonly assumed to be solid material spheres, imagined as minute billiard balls. Kelvin despised that idea. To him the material model for the atom was unsatisfactory as it offered no explanation for the fundamental properties of matter. He dismissed the billiard ball model as being superficial and naïve saying:

Fig 30 Lord Kelvin

"...the monstrous assumption of infinitely strong and infinitely rigid pieces of matter... Lucretius' atom does not explain any of the properties of matter..." [2]

With the endorsement of Lord Kelvin, the Helmholtz theory of the vortex atom replaced the solid particle theory for atoms in the last quarter of the nineteenth century. James Clerk Maxwell, who developed the equations for electromagnetic theory, spoke highly of it when he wrote for the Encyclopaedia Britannica:

> "...the vortex ring of Helmholtz...satisfies more of the conditions than any atom hitherto imagined..." [3]

J. J. Thomson, who discovered the electron, was also a proponent. He said:

> "...the vortex theory for matter is of a much more fundamental character than the ordinary solid particle theory." [4]

When I was sixteen I read about the Victorian theory for the vortex atom not in a physics book but in a book I borrowed from an antiquarian library, entitled, *An Advanced Course in Yogic Philosophy* [5] by William Atkinson, penning as Yogi Ramacharaka. The publication date was January 1905. I was galvanised. Yogis in Imperial India were not only centuries ahead of the British scientists but they had perceived the smallest particles of matter to be vortices of prana. I knew that prana was the Indian word for energy and that Einstein had published $E=mc^2$ later that same year. It struck me that if yogis could anticipate Einstein then their idea that fundamental particles of matter are vortices of energy could be valid. I was driven. I knew I had a life mission ahead of me to pick up the vortex and test it against modern physics. I had read somewhere that Einstein described matter as frozen light. I realised the description lacked true insight. The vortex offered a far more pertinent appraisal of how the dynamic state of energy could be contained in seemingly inert matter.

The Victorian physicists thought the atom was the fundamental particle of matter. They thought it was a vortex in the ether; they seemed to be unaware the ether had been be disproved and could not have known that in the twentieth century the atom would be reduced to subatomic particles. I realised it must have been the subatomic particles within the atom that yogis perceived as vortices of energy.

I was encouraged by my professor, Gareth Owen, Dean of the faculty of science at Queen's University of Belfast (the birthplace of Lord Kelvin) to develop a theory. I started work in earnest on a subatomic vortex theory during my first term there. However, its wasn't long before bang went the theory. My initial attempt to resurrect the vortex was quashed by a physics lecturer on the grounds that vortices have axes of spin whereas subatomic particles have no poles that reveal an axis of spin.

The problem was resolved in in a remarkable way. It was December 1968 and I had just turned twenty. During the Xmas recess I was billeted with my brothers

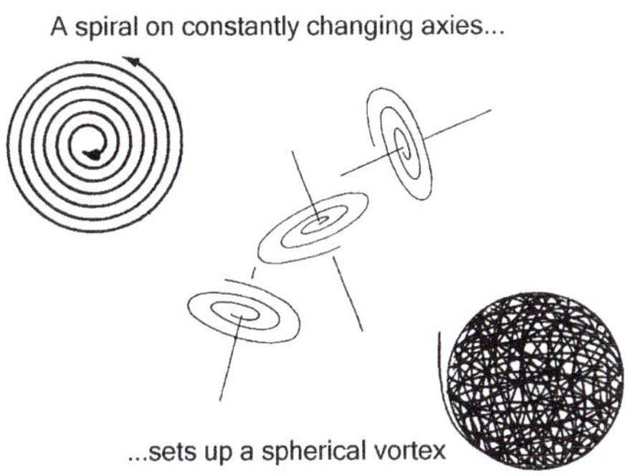

Fig 31 The ball vortex

and a friend of my father's, Harry Hanmer, in a caravan at Crackington Haven in Cornwall. Harry had asked me to hold up a hank of wool so he could wind it into a ball. As I watched him I was mesmerised.

I saw a ball vortex forming in his hands as he spun the wool on constantly changing axes. Watching the formation of the ball of wool I realised the subatomic particle might be a spherical vortex without apparent poles because the axis of spin of energy could be constantly changing. I was jubilant. I had a model, which I use to this day, to describe the way vortices of energy could form corpuscular subatomic particles of matter. In those few short moments watching Harry wind wool I saw the power of using simple models to describe quantum physics in a way ordinary people like myself could understand. Harry was enthralled by my description of how energy forms mass. From a ball of string the quantum vortex was born.

Fig 32 Gabriele Veneziano

The vortex theory is string theory. In 1969 the Italian physicist Gabriele Veneziano stumbled across the possibility of using a model of vibrating strings to describe fundamental particles in a set of equations he was working on. The vortex theory carries the idea that energy strings can spin as well as vibrate and spinning strings could account for the subatomic particles of matter.

I wondered if mass might be set up by the spin of energy in the vortex. Energy has no mass so laws pertaining to mass, like the law of conservation of angular momentum, would not apply to the spin of vortex energy forming a massive particle but mass could result from the vortex spin of energy because the mass of a particle is its ability to resist movement. This is called inertia. Richard Feynman said that the laws of inertia have no known origin [6] but the ball vortex provided a simple and straightforward way to account for the tendency of massive particles to stay put unless forced to move. Spin sets up inertia. This is demonstrated by a gyroscope. Gyroscopes can balance on a string because the spin of a gyroscope resists motion out of the plane of spin. This property is also depicted by a spinning pebble skipping across a pond. The spin of the pebble prevents the

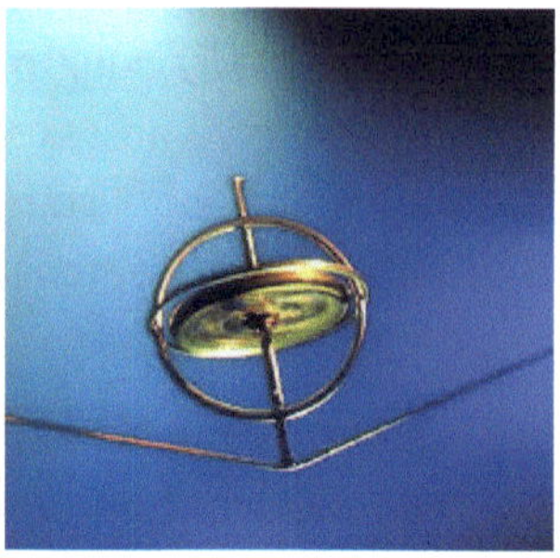

Fig 33 Gyroscope

pebble falling out of the plane of spin into the water. In a ball vortex the spin of energy on infinite planes, forming a spherical subatomic particle of matter, would set up resistance to its movement in any direction.

In the vortex theory the mass of a particle is defined as the amount of energy spinning in the vortex that forms it. Vast amounts of energy can be stored in mass. This is clear from $E=mc^2$. To appreciate how a vast amount of energy can be stored in the minute mass of a subatomic particle, imagine it as a ball of wool. Wool balls are very compact forms of wool. Likewise a ball vortex would be a very compact form of energy. As a store of energy the ball vortex can provide a simple explanation for potential energy in matter; the way energy can be stored in mass and then released as nuclear energy or gravitational energy.

A vortex is a three-dimensional spiral. The three-dimensional spin of energy in the ball vortices forming subatomic particles could confer three dimensionality on matter. This might explain why we live in a three-dimensional world. Three-dimensional extension is a fundamental property of the vortex. We can visualise quantum spin as a three dimensional spin system of energy. This and the fact that matter, space and the forces associated with matter extend in three dimensions adds strength to the vortex theory.

The vortex theory can be used to explain away materialism. Materialism is a philosophy based on the idea that substantial particles exist with irreducible properties and energy is the measure of their motion. In classical physics they were called atoms. In modern physics they are called quarks. Material substance can be defined in terms of the irreducible properties of inertial mass, potential energy and three-dimensional extension. In the theory of the vortex of energy these irreducible properties are reduced to spin. The philosophy of materialism can be explained away as spin in the double meaning of that word.

Albert Einstein established that energy is the prime reality. He dismissed the idea that material substance underlies everything. In his book, *Fields of Force: The Development of a World View from Faraday to Einstein,*[7] William Berkson made this clear when he wrote:

> "Einstein was difficult to understand... not because of his ideas or the mathematics he employed, but because of his world view. Einstein denied the substantiality of matter and the field, while maintaining their reality". [7]

Einstein established it is motion and not material substance that is fundamental to everything. This follows from $E=mc^2$ where he equated energy with mass. But what is energy? In physics, energy is defined as a measure of activity. Energy has no mass or substance so it appears to be pure activity without any medium of support. Greek philosophers invented the concepts of material atoms and fluid mediums as a substantial basis for movement otherwise they would be left with the dilemma of movement where nothing exists that moves. In the nineteenth century scientists presumed the existence of a luminous ether field permeating space to support the waves of light. That was disproved by Albert Michelson and Edward Morley in 1887 but in the 21st century this presumption has been resurrected in a different form. Physicists now believe the Higgs field permeates space as a supposed underlie to the excitation that gives rise to Higgs bosons. This is a step backwards because in the 20th century these materialistic notions were rejected by the more enlightened quantum theorists who came to accept that nothing underlies energy. Some overcame the dilemma of nothingness by presuming energy to be an act of consciousness rather than the action of a field or material substance. In his book *The Self-Aware Universe*,[8] Amit Goswami concluded that consciousness is the ground all being and in *The Mysterious Universe*:[9] the Cambridge astrophysicist Sir James Jeans wrote:

> "Today there is a wide measure of agreement that the stream of knowledge is heading toward a non-mechanical reality; the universe begins to look more like a great thought than a great machine. Mind no longer appears as an accidental intruder into the realm of matter; we are beginning to suspect that we ought rather to hail it as the creator and governor of the realm of matter."

If the Universe were a great thought it would not be beyond the bounds of possibility that there might be a great thinker. My supposition is that particles of energy are more like thoughts than things. I believe thoughts could set up a seemingly solid world through 3D quantum spin. Art transforms the abstract into tangible reality and as far as I am concerned the quantum vortex is an art form. The three-dimensional quantum spin of energy sets up a standing state that can transform the abstract into the apparent. This occurs because as energy travels in a curved path its velocity is less than when it moves in a straight line. If it moves in a spiral its straight-line velocity goes to zero. Energy spinning on a stationary point can set up a situation of stasis that gives rise to the illusion of substantiality as the physical properties of materiality. The yogis described this as 'maya', the illusion of forms. Countless trillions of

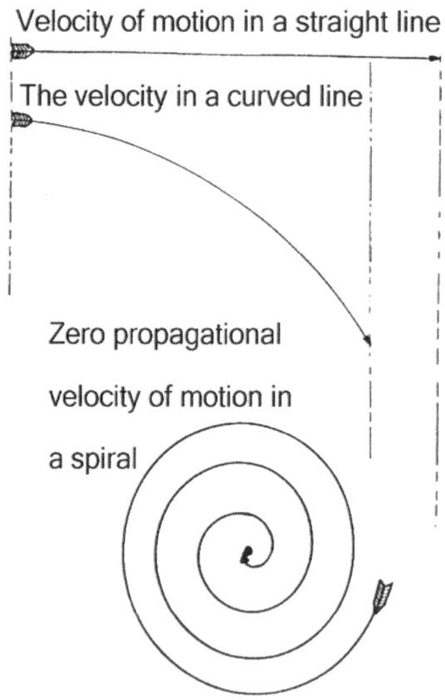

Fig 34 Spiral motion

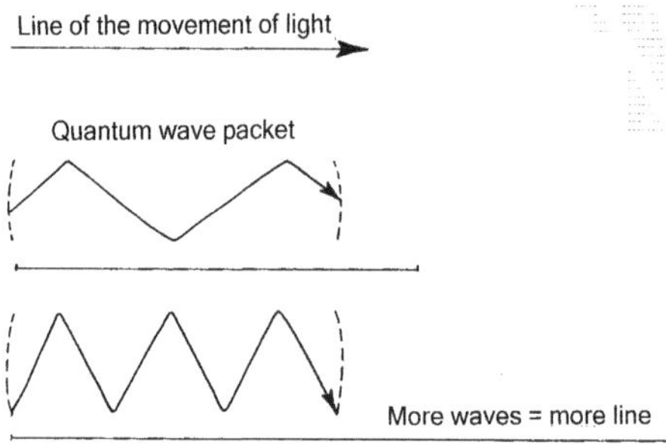

Fig 35 Wave packet

thought-form vortices overlapping and interacting with each other could set up apparently substantial worlds formed of the vortex force fields we perceive as atoms of matter. Out of these whirlpools of light, planets could appear and become populated by sentient beings like you and I, and the first nation Australians who recognised we are living in a dream state.

The other form of energy is the wave. Wave-propagation is the form of energy described as a quantum. This is the basis of light. A quantum particle of light is also called a photon. If we imagine a particle of light as a bundle of vibrating string, a higher frequency represents a greater number of waves in the bundle. This would result in a greater length of string. If the length of string is taken to represent the energy contained by the bundle then immediately we can understand why the energy in a quantum is proportional to its frequency of vibration. The constant of proportionality is called Planck's constant, after Max Planck, and it is symbolised by the letter 'h'.

Physicists have endeavoured to explain the Universe entirely in terms of the wave form of energy. I believe this is just half the story. My contribution to physics is to suggest there might be another form of energy that has been overlooked in modern physics. The vortex. To make way for this second form of energy I have to redefine terms. I define the term quantum as a particle of energy that can occur either in a wave form or a vortex form. I define inertia as a property of energy set up by either of these forms of motion. The propagational wave-form quantum sets up a system of kinetic inertia, which keeps moving unless it is stopped. The static vortex-form quantum sets up a system of static inertia which stays put unless it is moved. Thus the wave and the vortex forms of energy set up opposite inertias.

A good analogy for inertia is the behaviour of children. In the evening their inertia is kinetic. It is hard to get them to stop playing and go to bed. In the morning their inertia is static. It is hard to rouse them from bed, get them dressed and off to school. They are the same little bundles of energy which in different situations have different inertias.

If energy is pure movement should the movement cease the energy would cease to be and the Universe would vanish. In science this is the basis of the first law of thermodynamics that energy is neither created nor destroyed. Inertia embodies this creed. The common concept of inertia as stasis is an illusion set up by spin. In reality the inertia of energy is unstoppable motion. Quantum particles of energy, in the form of wave or vortex, appear as particles of perpetual motion where nothing exists that moves. As prime reality they seem more like acts of imagination than material things. That is quantum reality.

As well as the form of its motion an underlying feature of a quantum of energy is the speed of its motion. In his special theory of relativity Einstein determined that everything in the world is relative to the speed of light. This could be because the speed of motion in every quantum of energy in the physical world is the common speed of light.

The propagating wave-quantum of energy has a lot in common with thought because in its vibration it can transmit information. The static vortex-quantum has a lot in common with memory because it can store information. Considering particles of energy as thoughts and the Universe as a mind is more in the realm of mysticism than physics but Wolfgang Pauli's professor and lifelong mentor, Arnold Sommerfeld, impressed on his students that science emerged from mysticism and has never completely separated itself.[10] Mysticism shines through quantum physics so it should not come as a surprise to see a reintroduction of the vortex into physics coming from Indian mysticism. Nonetheless the vortex does have provenance in physics. The vortex atom theory was not a bad idea but it was applied to the wrong thing. The vortex theory was applied to atoms and not to subatomic particles. That was not the fault of the Victorian physicists. In their day subatomic particles weren't yet discovered. The atom was thought to be the smallest particle of matter but that unfortunate error led to the abandonment of the vortex theory when it was unable to account for many new discoveries about the atom including its colour spectra.

Niels Bohr succeeded in explaining spectral phenomena by applying Einstein's quantum theory to the atom. He suggested the atom was a system of electrons, in distinct orbits, that could move into higher orbits in a quantum leap by absorbing energy. Bohr explained that the coloured spectra of atoms represented photons of light emitted by electrons when they released energy as they fell back to their original lower orbits. He said the colour of the light emitted represented the frequency difference between the higher and lower orbits. Bohr's success in explaining the line and colour spectra of atoms enabled him to shift the centre of physics from Britain to Europe where quantum theory then quantum mechanics had become the major theories in physics.

The gentlemen physicists of Britain played cricket by the rules. When the British physicists realised that their theory for the vortex atom couldn't explain discoveries about the atom, especially spectra, they accepted they had been bowled out. The European team then took its innings under the captaincy of Einstein who handed the cap to Bohr. Heisenberg was in to bat with his quantum mechanics until he was bowled out by Chadwick in the British team. When Chadwick discovered the neutron, in 1932, Heisenberg should have finished his innings. But the Europeans, unfamiliar with the rules of cricket, continued to bat despite the fact that the uncertain wicket of quantum mechanics has been knocked out by the certainty of an electron in the neutron cricket ball. The time has come for the British team to declare the Europeans out and take an innings with a revised version of the vortex theory applied at the subatomic level.

1. Isaacson W. *Einstein: His Life and Universe*, 466-468, Simon & Schuster, 2007
2. Thomson W. *Popular Lectures and Addresses* 1841
3. Maxwell J.C., *Encyclopaedia Britannica* 1875.
4. Thomson J.J. *Treatise on the Motion of Vortex Rings*, Cambridge University, 1884
5. Yogi Ramacharaka, *An Advanced Course in Yogic Philosophy*, The Yogi Publication Society, January 1st 1905
6. Feynman R. *The Character of Physical Law*, Penguin 1992
7. Berkson W. *Fields of Force: The Development of a world view from Faraday to Einstein,* John Wiley & Sons, 1974
8. Goswami A. *The Self-Aware Universe,* Putnam, 1995
9. Jeans J. *The Mysterious Universe*, Cambridge University Press, 1930

5. In My Beginning Is My End - Shane Guffogg

In this intriguing image I see the endless spinning strings of energy - Ash

ESSAY 5

The New Vortex Theory

"A future theory must bring about a deep unification of foundations."

- Wolfgang Pauli

Précis: When he was five, in bed recovering from flu, Albert Einstein's father gave him a compass to play with. He was fascinated by the way the needle always faced the same direction no matter how he turned the compass. Not yet blinded by science he deduced, in his young mind, that space was gripping the needle. The idea that space was something real that could act on things like compass needles never left him. Later it would give rise to his theories of relativity.

The vortex theory supports Einstein's view of space as a real form of energy rather than an empty void. In the vortex theory space is viewed as the infinite extension of vortex energy. If matter is the dense vortex energy we perceive, space would be the sparse vortex energy stretching beyond our perception. As matter is particulate so would be space. Like an ocean composed of drops of water, so space would consist of countless particles or bubbles of space, each existing relative to the other. Every vortex could provide the space in which another vortex would be the movement and vice versa. This could account for Einstein's special theory of relativity, expressed as the universal law: 'love thy neighbour as thyself'.

In the vortex theory, the forces associated with matter are not treated as particles or fields acting in space, they are treated as actions of vortex energy perceived as empty space. Each particle of space overlaps every other. As extending vortices of energy they are dynamic systems that can act, each one against another. As such, vortices of energy could be acting as force fields. Each vortex of energy, through its extension as space, could be a particle of electric, magnetic and gravitational force. Space is not electrically active because there appear to be an equal number of opposite charges in existence, which cancel each other out. Nonetheless, if the extensions of rotating vortices of energy set up magnetism then, as extending vortex energy, space could be magnetically active. In that respect the little Einstein was right.

Einstein suggested gravity is caused by a distortion of spacetime by mass. In the vortex theory space is thought to be curved, not because it is distorted by mass but because it is an extension of mass. The vortex energy extending from all the particles of matter in the Sun could set up vast concentric spheres of space extending from the Sun. Starlight passing by the Sun could then follow the curvature of the Sun's space much as a car would follow the curved road around a roundabout. In the vortex theory the cause of gravity is different to what Einstein believed.

In the vortex theory, all the vortex energy in matter is thought to extend out into a maximum size sphere of space through which it flows into a world of antimatter. At the other end of the size spectrum of space, a reverse flow of vortex energy passing through a point of singularity in the centre of every particle of matter could form a particle of antimatter beyond it. If that were so, every form and action in the world of matter would be replicated in the world of antimatter. The opposite directions of flow of vortex energy through the worlds of matter and antimatter, represented as opposite positive and negative charges, would set up the symmetry of mirror worlds.

Between matter and antimatter there is an electric force of attraction. Every body of matter in and on the Earth could be pulled down by a force of attraction coming up through the centre of the Earth 'from the antimatter Earth beyond it' as they share a common centre. This could account for gravity. If the electric force is an interaction acting from the outside of the perceived aspect of the vortex particle, the gravitational force could be an electric force of interaction acting from inside the vortex particle. Einstein was unable to unify his 'spacetime distortion' theory for gravity with the electric force, but in the vortex theory the force of gravity is unified with the electric force because they are the same thing acting through different regions of space. If electric charge and gravity are both vortex interactions then the vortex of energy could be the unified field Einstein searched for to the end of his life.

~

Fig 36 Albert Einstein

Essay: In *The Evolution of Physics*[1] Einstein wrote:

"There is no sense in regarding matter and field as two qualities quite different from each other... could we not reject the concept of matter and build a pure field physics? We could regard matter as the regions...where the field is extremely strong."

Albert Einstein was convinced that matter and the field could be thought of as strong and weak regions of the same thing. If the vortex of energy is treated as a quantum particle field, then it becomes clear that matter may be the dense central region of the field of vortex energy – where it is strong – and space may be the sparse peripheral region of the vortex energy field – where it is weak – and forces could be the effects of the 'particulate' quantum fields of vortex energy, where one vortex overlaps and interacts with another.

The quantum field of vortex energy is illustrated by the story from India of a group of four blind men who came upon a tethered elephant. One felt its leg and said it was a tree. Another felt its ear and said it was a leaf. A third felt the elephant's trunk and insisted it was a python, while the fourth gripped its tail and said it was a snake. Blind as they were to the elephant they thought it was different things. Blind as we are to the vortex of energy as one thing with different effects, we have described it as many different things.

If the vortex of energy is treated as a unified field of matter, space and forces then, like the elephant in the story, we have given it different names according to different ways we perceive it. In the vortex theory, matter is treated as dense space and space as sparse matter; and the forces of electric charge, magnetism and gravity are considered to be the dynamic effects of one vortex of energy on another. According to the vortex theory, the action of space as a force is apparent in gravitation but not as an electric force. This is because there are an equal number of opposite electric charges in the Universe, which cancel out each other's effects. That is why space is electrically neutral. This conclusion is taken from the Law of Conservation of Charge pronounced by Lord Rutherford that for every positive electric charge in the Universe there exists an equal but opposite negative electric charge.

The fact that magnetism is an action of space was apparent to Albert Einstein when he was five. He was in bed recovering from flu and his father gave him a compass to play with. He was fascinated by the way the needle always faced the same direction no matter how he turned the compass. He was not yet blinded by science and deduced, in his young mind, that space was somehow gripping the needle. The idea that space was something real that could act on things like compass needles never left him. Later in his life it led to his theories of relativity.

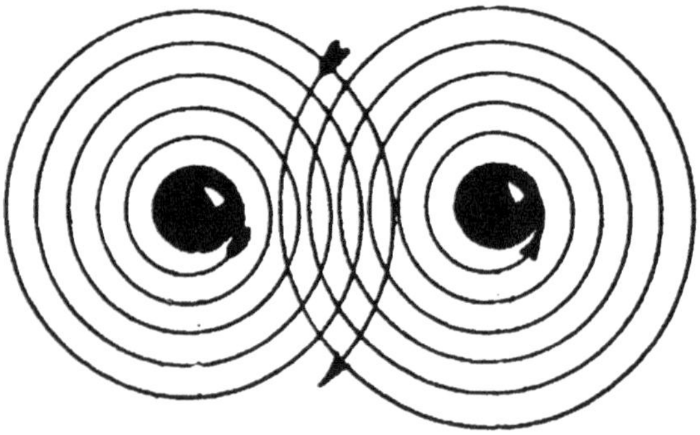

Fig 37 Vortex interactions

In the vortex theory, the interactions between extending vortices of energy are thought to account for the forces associated with particles of matter because each vortex of energy is a dynamic system capable of interaction. When they overlap, each vortex of energy would be a real force-carrying particle relative to every other particle of vortex energy so there would no further need for the arbitrary assumption of virtual force-carrying particles. Real particles existing as overlapping vortices of real energy would be carrying a real force through their real interactions.

The overlap of vortices of energy requires extension. The infinite extension of electric, magnetic and gravitational forces suggests that vortices of energy are infinite extensions. It is obvious that a vortex of energy must be an unlimited extension because energy is neither created nor destroyed. As the vortex energy extends out its intensity would diminish but never vanish altogether. This is illustrated by the principle that one can never get rid of a pie by dividing it into ever smaller pieces; it simply gets more dispersed. Like the pie, the extending vortex energy would never cease to be, it would simply spread out more thinly.

Three-dimensional extension is a fundamental property of the vortex. People take it for granted that we live in a three-dimensional world. The vortex theory explains why this is so. The fact that the forces of gravity, electric charge and magnetism extend in three dimensions supports the idea that they are extensions of vortex energy. The three dimensional extensions of matter, space, and forces adds strength to the idea they are all manifestations of vortex energy. They are just the different ways we perceive it.

As a three-dimensional extension, the vortex of energy expands according to the inverse square law. This means the intensity of energy at a distance would diminish to a quarter as the distance doubles and to a nineth as the distance

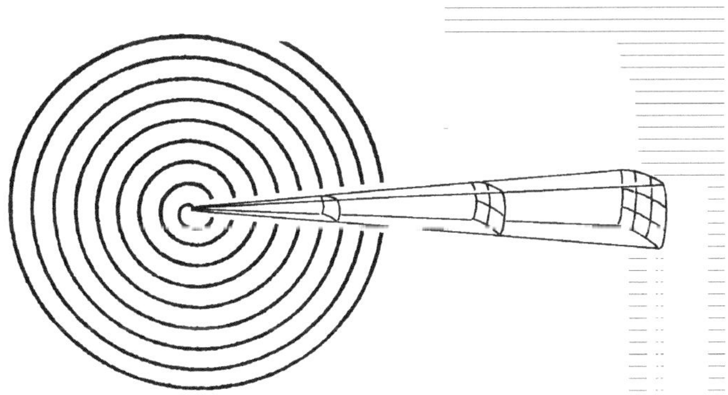

Fig 38 Inverse square law

trebles. If the strength of vortex interactions is determined by the intensity of vortex energy interacting, then the force of interaction between overlapping vortices would be expected to obey this law. The forces of electric charge, magnetism and gravity do in fact obey the inverse square law in their strength of interactions. This supports the premise that they are caused by vortices of energy interacting.

Subatomic particles act at a distance instantaneously through the forces of electric charge, magnetism and gravity associated with them. Einstein called the ability of particles of energy to act at a distance "spooky" and he strongly resisted the idea of quantum entanglement.[2] But experimental physics has come out in support of the quantum entanglement of photons of light, and the infinite extension of quantum vortices of energy suggests the propensity of subatomic particles to action at a distance through quantum vortex entanglement. If subatomic particles are infinitely-extending quantum vortices of energy they would be in a state of constant entanglement or overlap and so instantaneous interactions between them would be possible. This we observe in the forces of electric charge, magnetism and gravity.

If space is an aspect of the vortex of energy, then space would be quantised or particulate. A quantum of space could be described as a force-carrying particle if the forces associated with subatomic particles are accounted for in terms of vortex interactions. Overlapping spheres of infinitely-extending quantum vortices of energy could account for the overlapping spheres depicted by the ancients in an image called the flower of life. It was believed that the flower of life is a symbol that holds hidden knowledge of the Universe.

Fig 39 Flower of Life

To appreciate the vortex theory for electric interactions imagine you are in a quantum vortex and experience it as a fireball. As you move outwards from the fiery centre, the energy rapidly diminishes. Suddenly it vanishes altogether as though you have come to the end of the vortex and broken out of its fiery domain. However, the apparent surface of your vortex is not its boundary, it is merely the last intensity of energy that you can perceive. Looking out from this point into the darkness you see other fiery vortices moving about. You assume that all these spinning balls of light are separate vortices occupying a void of darkness – like stars in the night sky. However, your senses deceive you. The apparent void of darkness is full of energy extending from all the vortices. As the invisible energy from one quantum vortex overlaps that of another there is an interaction between them. Standing on your vortex and seeing nothing but a void between you and the others you are perplexed at the inexplicable attractions and repulsions between them and speak of spooky 'action at a distance'.

In physics forces are considered to be carried by fields. These are described as 'force fields.' In the vortex theory the arbitrary assumption of extraneous force fields is rejected in favour of intrinsic force fields embodied in the interactions of extending vortices of energy.

In the vortex theory, electric charge is thought to be set up by the flow of vortex energy growing or shrinking in or out of a subatomic particle. Magnetism is considered to be a consequence of the rotation of the spheres of vortex energy extending from a particle. This accounts for the natural magnetism set up by spinning charged particles in atoms. Most atoms are not magnetic because the electron vortices in them occur in pairs spinning in opposite directions, which causes a cancellation of the magnetic effect. An atom like iron is naturally magnetic only because it contains an unpaired electron.

The quantum vortex enables us to understand why electric charges occur as whole or integer units. Each quantum vortex contributes a single whole unit

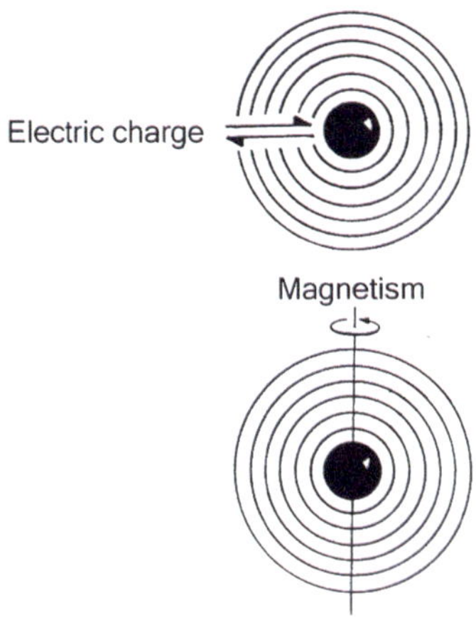

Fig 40 Charge and magnetism

of charge, which is why there are no fractional charges in nature. An analogy of soldiers can help us understand this. Armies are made up of single whole soldiers. These represent unitary charges. Each soldier has a rifle. The size of the soldier doesn't matter. Whether the soldier is big or little he is equally effective as any other soldier with his rifle. That explains why the charge associated with a quantum vortex is independent of its mass. Fractional soldiers without a limb are either discharged or not admitted into an army. Fractional rifles without a stock or a barrel would not work. In like manner the fractional charges of quark theory may work in theory but they are a fiction because they are not observed in nature.

Fig 41 The soldier analogy

The vortex of energy upholds Newton's laws of motion. Newton said, "For every action there is an equal and opposite reaction". Each extending vortex could be acting as a force against which another is a reaction. Each vortex of energy would exist and act relative to every other vortex. The vortex theory points to reconciliation between Newton and Einstein because Newton's laws can help us understand Einstein's theory of relativity.

In his special theory of relativity Einstein dismissed the concept of space existing as an absolute void. This sets up a conundrum. How is it possible to have motion without space?

Einstein overcame this dilemma by localising space and introducing relativity. The vortex theory can help us visualise this. If an apparent void of 'local' space is set up by an extending vortex of energy then the motion of another vortex or a photon of light could occur relative to it. One vortex of energy could be an extension of space relative to which another vortex would be a form of motion. Each vortex existing as an extension of space relative to a neighbouring vortex or wandering photon, acting as a system of motion, would illustrate Einstein's theory of relativity and present 'love thy neighbour as thyself' as a universal law.

In *Einstein: His Life and Times* [3] I read that when a reporter asked him to put his theory of relativity in a single sentence, Einstein replied that if you remove matter from the Universe you also remove space and time. Einstein's remark affirmed the view that matter and space are connected. That would make sense if they were both vortex energy. Before I read the book I had ascertained from its three-dimensional extension that space could be vortex energy extending beyond perception from matter into infinity. I concluded that matter and space were not different things; they were just different ways we perceive the same thing. Just as the form of a vortex in a stream is maintained by the constant flow of water, so the forms of matter and space would be set up by the constant flow of energy in the vortex. Consistent levels of energy intensity at every increment of distance from the centre of the vortex would set up stable forms in matter and shells of space through which the vortex

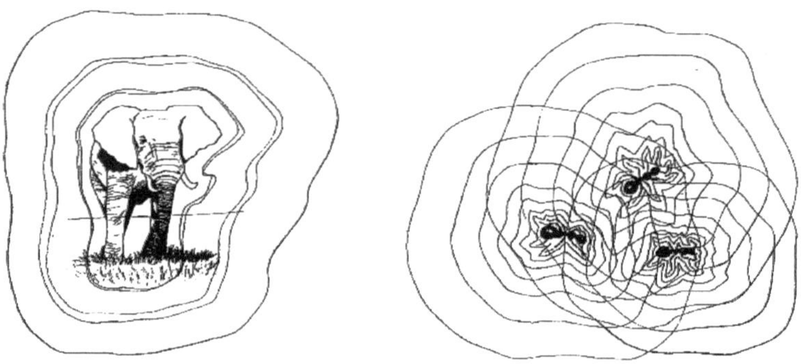

Fig 42 Space bubbles

energy would constantly flow. That reminded me of the yogic description for the delusion of materiality as maya, the illusion of forms.

According to the vortex theory, every body of matter is surrounded by its own extending quantum field or 'bubble' of space and the shape of this space would be an aura extension of the shape of the body of matter it extends from. The moving space extending from an underground stream or blind-spring might deflect diviners' rods. That could explain the ancient art for detecting water under land. You can detect the aura of space extending from a table by swinging a pendulum alongside it. When you come to a corner you will notice the pendulum change direction and swing round the corner as if it is tracking out the contours of the space extending from the table. You can repeat this demonstration with any body of matter.

I used the idea of space being an extension of matter to explain the central premise in Einstein's special theory of relativity, that the measured speed of

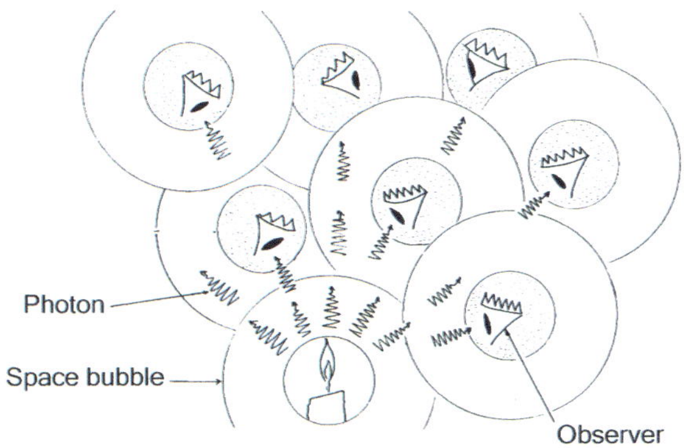

Fig 43 Special Relativity

light is independent of the velocity of the observer. It was obvious to me that if we were to measure the speed of light, the photons whose speed we were measuring would be in our own bubble of space, which would move with us as we move. Measuring the speed of light relative to our own local space would cause the measure to be independent of our own movement. Michelson and Morley found that the speed of light measured on Earth is independent of the movement of the Earth. That could be taken to infer that the Earth's local space moves with the Earth.

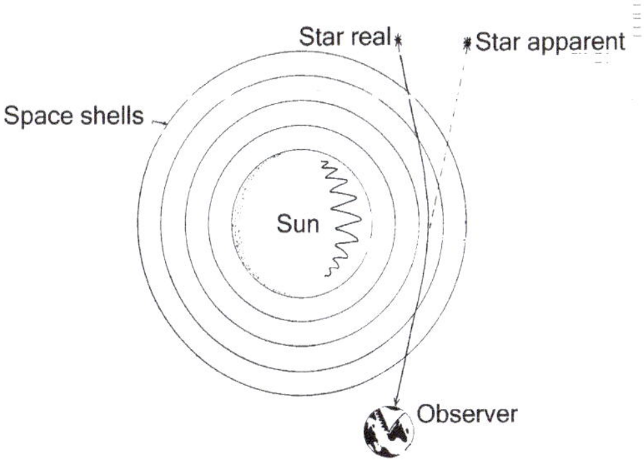

Fig 44 Curved space extending from the Sun

If the space extending from every body of matter is an extension of the shape of that body then the space extending from the Sun would take the form of concentric spheres. Starlight passing close to the Sun would follow the curvature of the Sun's space as surely as a car follows the road going round a roundabout. This could account for the deflection of starlight predicted by Albert Einstein in his general theory of relativity, which was confirmed by the Royal Society in 1919.

Einstein accounted for gravity by assuming space and time are a four-dimensional spacetime continuum distorted by mass. That was an arbitrary assumption which can be discarded because in the vortex theory space curvature is

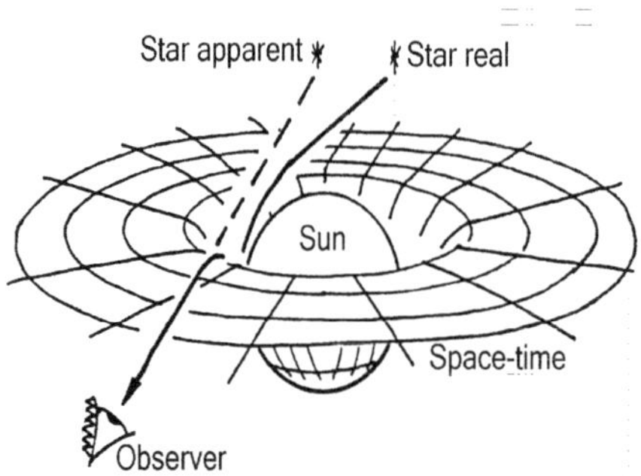

Fig 45 Einstein's spacetime distortion theory

thought to be caused by an extension of mass not a distortion by mass and gravity is shown to have a different cause to what Einstein thought because his thinking on gravity preceded the discovery of antimatter.

In 1931, Paul Dirac predicted the existence of antimatter in the math he was developing for quantum mechanics. It seemed like science fiction until later that same year Carl Anderson at the California Institute of Technology spotted positively charged electrons emerging from a thin lead sheet he was bombarding with gamma rays.[4] Dirac called them 'positrons'. I accounted for this experiment as two wave trains of energy in a gamma ray photon driving through the nuclear vortices in an atom of lead. As they did so they were transformed into a pair of ball vortices with equal energy spinning in opposite direction. This would explain the electron-positron pair particles with equal mass but opposite sign of charge. When the positron subsequently encountered an electron their mutual mass was annihilated and they reverted back to gamma ray photons. I visualised this as two equal but opposite vortices of energy unzipping one another, enabling the energy to revert from the vortex form back to the wave form again.

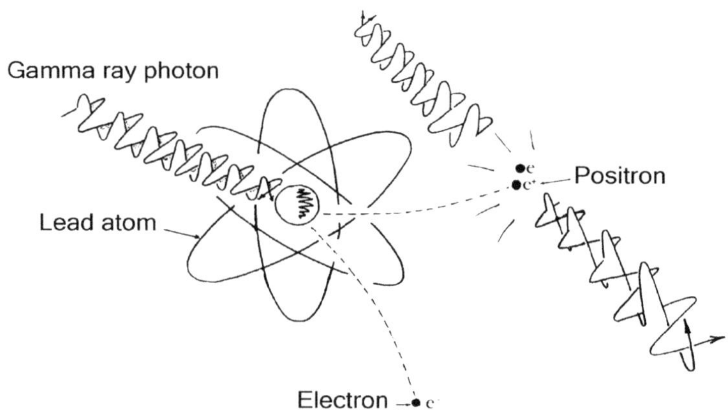

Fig 46 Antimatter creation and annihilation

In the 1950's Wolfgang Pauli was exploring the math of antimatter and came to an astonishing conclusion. To quote Arthur Miller from *137: Jung, Pauli and the Pursuit of a Scientific Obsession:*

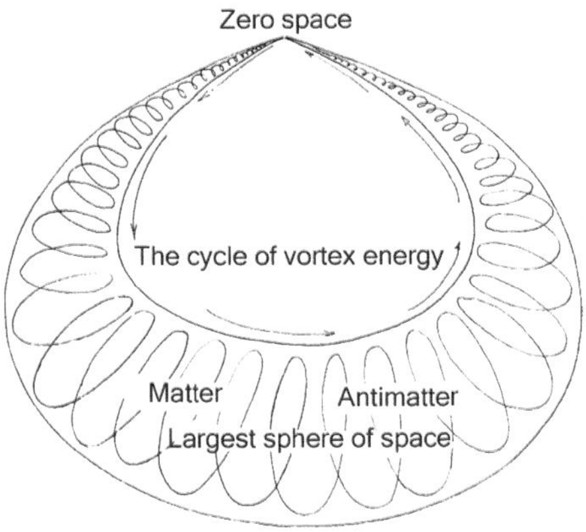

Fig 47 Universal cycle of vortex energy

"The year was 1952...Pauli began his investigation into mirror symmetry (parity) by looking at time reversal...By the mid1930s physicists agreed that every charged particle must have a matching antiparticle...In 1954...Pauli decided to look deeper into the whole subject. Instead of exploring charge conjunction (C - replacing particles with antiparticles in equations), parity (P), and time reversal (T) separately, he looked at all three together- the whole combined operation of CPT makes the astonishing assertion that a mirror universe - in which all matter is replaced with antimatter, all positions are their own reflections, and even time runs backwards so that all speeds are reversed - would actually be indistinguishable from our own." [5]

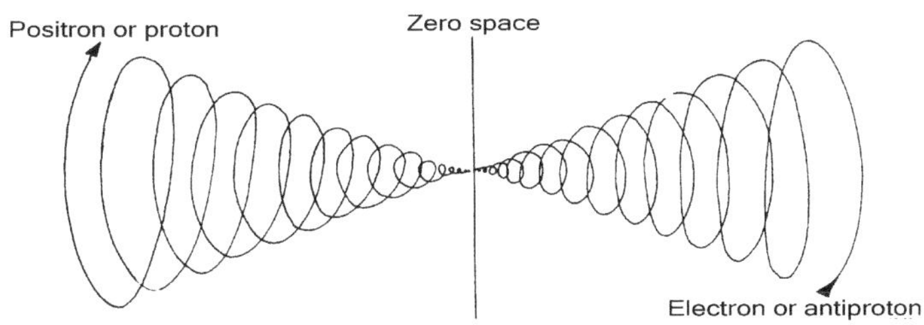

Fig 48 Antimatter matter funnel

Unaware of Pauli's discovery I came to the same conclusion but by a different route. It began in the summer of 1969 when I was sitting by a stream in Cornwall pondering on where the energy spinning in the subatomic ball vortex came from and where it went. The answer that came to me was just as in the water cycle, water in a stream circulates endlessly between rain drops and oceans so vortex energy could circulate endlessly between vortex particles and swirling galaxies in matter and antimatter, through two halves of the Universe.

I imagined energy spinning into the centre of an electron vortex as though it were a funnel. It then passed through a point of singularity at the centre, as though through a tunnel, and then spinning out the other side it would form a positron vortex with equal mass but positive charge. From there the vortex energy would expand through antimatter atom, planet and galaxy until it reached a frontier of maximum size after which it would contract again through galaxy, planet and atom in matter, then shrinking back through the electron with its negative charge it would pass through singularity and begin the cycle all over again.

Fig 49 The photon

The cycle of vortex energy through every vortex in matter into an equal but opposite vortex of antimatter would mean that beyond the centre of every subatomic particle of matter there would be an identical particle of antimatter. If that were so every form and action in the world of matter would be faithfully replicated in an identical, mirror world of antimatter. This realisation fitted Pauli's discovery precisely. For the vortex theory to work there would have to be a mirror you in antimatter beyond the centre of you in matter, perusing these words on a mirror page in a mirror antimatter book and the two of you would be sharing the same photons of light to read by.

Every photon of light contains two transverse wave trains of energy travelling at right angles. These were thought to be electric and magnetic fields but that cannot be true because magnetic and electric fields are both effective in the world of matter, whereas in fluorescence and polarisation it has been shown that only one of the fields in the photon reacts with matter, and also, in photography, only one field of energy in light reacts with the photographic plate. The other field appears to play no part in the world of matter.

This anomaly is reflected in quantum mechanics, where only half a wave quantum isf effective in quantum mechanical events. The basic premise of quantum theory is that energy is particulate and events in the quantum world should occur as integer multiples or whole numbers of quantum particles of energy. The energy in each wave form quantum is equal to its frequency times Planck's constant, which is symbolised by the letter h. But quantum mechanics reveals that natural particles like electrons and protons are governed by half Planck's constant, represented by ½h. In physics ½h is given as the value of quantum spin. This tells us that only half of each particle of energy is involved in events we observe in quantum mechanics. Could it be the other half is being observed in the antimatter world?

Carl Anderson's experiments demonstrated that one field of energy in the gamma ray photon appeared to be presumptive matter and the other was presumptive antimatter. The vortex theory can only work if, at a quantum level, energetic events involve half a wave form quantum represented by half Planck's constant, symbolised by ½h. This is because only half of each quantum of waves can be driving a quantum mechanical event in the world of matter. The other half would be driving an identical, mirror, quantum mechanical event in the world of antimatter. Only then could every event in the world of matter could be replicated faithfully in the world of antimatter. The need I had to explain where the energy comes from and where it goes to in the universal cycle of vortex energy led me to this conclusion.

Subatomic particles of matter such as electrons and protons are each governed by ½h. In the vortex theory, ½h denotes a single string or line of the movement of light vibrating to form half a wave quantum or spinning to form a subatomic quantum vortex in matter. The other half of the quantum, ½h, would be vibrating in the world of antimatter or spinning to form a subatomic vortex in antimatter. This is how every form and action in the world of matter is replicated in the world of antimatter. The antimatter half of the universe mirrors every form and action in the world of matter. This was discovered by Pauli and is central to the vortex theory. The fact that quantum events in our world of matter occur as integer multiples of half Planck's constant, ½h, with the other ½h driving the same quantum mechanical events in antimatter reflects the mirror symmetry of the Universe. In the vortex theory every quantum of energy, either in the form of a wave train or paired vortex particles, is simultaneously in the world of matter and antimatter. I call this quantum loop symmetry.

Fig 50 The quantum loop

In his equations Dirac discovered that time appears to run backwards in antimatter. This can be explained in terms of quantum loop symmetry. If the direction of time, past-present-future is set by the direction of flow of energy in the quantum of waves, reverse time would suggest that the two lines of energy are flowing in a wave quantum or photon in reverse direction. That would allow for an endless circulation of energy in a wave form quantum. Dirac's discovery suggests time is circular and energy flows endlessly in circles.

To reiterate, what appears to be two strings of energy flowing in waves in the quantum may be a single continuous looped string of energy flowing backwards and forwards as two perpendicular wave trains forming a photon. Planck's constant could define a quantum loop of energy where one half of the energy loop exists in the world of matter and the other half in the world of antimatter. These two halves of the energy loop could occur either as a centre connected particle and anti-particle paired in the three-dimensional extension of matter and antimatter, or as an opposite directional flow of energy in the quantum looped two dimensional transverse wave trains in a photon. This is not easy to visualise but matter and antimatter exist in mirror symmetry so, in the mirror symmetrical worlds of matter and antimatter, time and events would be actually running in the same forward direction of past-present-future. It is only when they are viewed relative to each other that they appear to be in reverse direction; much as things appear reversed in a mirror. Whereas the experience of linear time would be the effect of half the quantum loop in either matter or antimatter, represented by ½h, the entire quantum loop would be circular time, represented by Planck's constant h, existing in the eternal now.

Between matter and antimatter there is an electric force of attraction which causes bodies of matter and antimatter to accelerate together. According to quantum loop symmetry, in the universal cycle of vortex energy, every form and action in the world of matter is mirrored in the world of antimatter. If that is so there would be an antimatter Earth beyond the centre of the matter Earth and there should be an accelerating force of attraction from the antimatter Earth acting through the centre of the Earth on every body of matter in and on the Earth and vice versa. This attraction would not be between a body in matter and its twin body in antimatter, it would between the body and every other body in antimatter. This vortex interaction, operating through the centre of the Earth, would cause a centralising pull on everyone and everything resulting in tug or a fall toward the centre of the Earth. This would be experienced simultaneously in antimatter toward the centre of the antimatter Earth. Observed in the fall of an apple, this effect was described by Sir Isaac Newton as gravity.

The theory of the quantum vortex presents an alternative to Einstein's theory for gravity as expressed in his general theory of relativity. By presenting gravity as an electric force of attraction between matter and antimatter, the vortex theory unifies gravity with the force of electric charge. This is what Einstein wanted to achieve. The vortex account for gravity suggests that the vortex of energy may be the unified field Einstein was searching for to the end of his life. Richard Feynman said that gravitation is, so far, not understandable in terms of other phenomena but In the vortex theory gravity is understandable

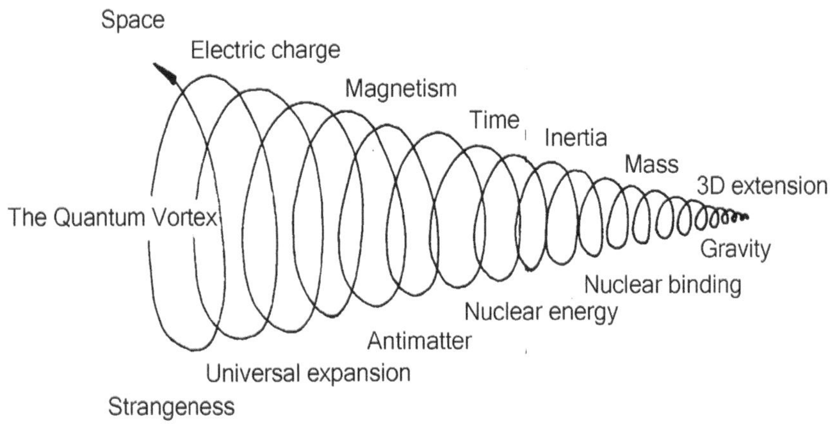

Fig 51 The unified field of vortex energy

in terms of practically all other phenomena associated with matter. The vortex theory unifies quantum spin, inertial mass, three-dimensional extension, matter and antimatter, space, electric charge, magnetism and gravity in the single quantum-field of vortex energy.

For a theory to be acceptable in science it must account for most things with the minimum of assumptions. The vortex theory does that. But it must also predict the outcome of future observations or scientific experiments. The vortex theory has also achieved that.

When I was considering where energy goes to, as the vortex extends into infinity, it became obvious to me that it was the size of space and not time, as Einstein supposed, that was the fourth dimension. We all travel in the fourth dimension of space when we grow from a single cell into a full-grown adult. I named this the 'Alician Dimension' after Alice in Wonderland because, in a story that anticipated the mirror symmetry of matter and antimatter, Alice shrank into a looking glass world. In the vortex theory the worlds of matter and antimatter are connected in both extremes of smallness and bigness. Symmetry would require an electric force of attraction to cause acceleration between matter and antimatter through the largest as well as the smallest spheres of space. So in addition to the gravitation attraction between matter and antimatter in smallest space, there should be an accelerating pull from antimatter coming into our world of matter through the largest size sphere of space connecting the two halves of the Universe. It would be the polar opposite to gravity. This pull would operate on the largest things; it would cause galaxies to accelerate toward the largest sphere of space and the closer they would be to it the faster they would accelerate toward it. This has always been a major prediction in the vortex theory. In the early 1970's I realised that the galaxies furthest from us would be accelerating faster from us than those closer to us. I also realised that quasars, then observed as the outermost and brightest things in space, could be stars undergoing annihilation over the largest sphere of space. Annihilation was obvious to me because quasars put out about 200x the energy of a star. Because stars are fuelled by nuclear fusion, which uses just over 0.5% of proton mass, anything generating 200x the energy of fusion must be using up 100% proton mass and the only process we know that can do that is matter-antimatter annihilation. I predicted the way galaxies accelerate from us in a lecture I gave to young members and guests at the Royal Institution of Great Britain, on 15 January 1975. I then included it in the Vortex Cosmology chapter of my book published in 1995.[6] The acceleration away from us of furthermost galaxies was verified subsequently by Saul Perlmutter in December 1997.[7] He spoke of this apparent accelerating expansion of the Universe as Dark Energy because he was in the dark about what caused it. He received the 2011 Nobel Prize in physics for his achievement. My satisfaction was to know that the vortex hypothesis qualified as a theory as it satisfied the requirements of the scientific method for a good scientific theory as outlined in *A Brief History of Time*, [8] by Stephen Hawking:

> "A theory is a good theory if it satisfies two requirements: It must accurately describe a large class of observations on the basis of a model that contains only a few arbitrary elements and it must make definite predictions about the results of future observations ."

As far as I am aware the vortex theory is the only string theory that has managed a prediction that has been successfully verified. Another achievement has been to provide a visualisation of quantum spin, which could reinstate realism in quantum physics.

The story of quantum spin began when Werner Heisenberg, then only nineteen, was tackling an equation Wolfgang Pauli couldn't solve. He was successful because in a flash of genius he rewrote the equation using half numbers and it worked perfectly. That was how the half quantum entered quantum mechanics. Bohr and Pauli were stunned. Heisenberg wrote back to Pauli, "Success sanctifies the means." Meanwhile, Pauli came up with his own stroke of genius. Bohr had applied three quantum numbers, representing the three dimensions of space, to electrons in atoms but Pauli, studying the scientific data, realised there had to be a fourth quantum number representing a fourth dimension. Then the twenty year old Ralph Kronig spotted something Pauli had missed. Pauli's number had the properties of spin along with Heisenberg's half quantum value. Thus quantum spin with ½h was born. To begin with it was thought to represent the top-like spin of an electron but physicists, George Uhlenbeck and Samuel Goudsmit, revealed that was not so. Quantum spin was a different sort of spin altogether, which nobody could visualise. That led Pauli to demand that realism, in the form of visual images, should be banished from quantum mechanics. To quote Arthur Miller from his book 137:

> "Spin was undeniably a property of an electron but it was entirely impossible to visualize it...Scientists had to accept that the fourth quantum number had no accompanying visual image. It was time for atomic physics to move on from trying to visualise everything in images relating to the world in which we live." [9]

If quantum spin is visualised as the vortex spin of energy forming an electron rather than the spin of the electron itself then the +½h and -½h quantum spin values could be applied to the direction of spin of energy in or out of the ball vortex instead of to the rotational directions of electron spin. ½h could indicate it is a single energy string in spin. If Bohr's three quantum numbers were applied to the three dimensions of spin in the ball vortex then Pauli's fourth quantum number could refer to the simultaneous spin of energy in the three dimensions, either in or out of the 'spherical' vortex centre. That would establish the fourth dimension as the dimension of size of space, which we experience every moment in the inhalation and exhalation of the breath.

1. Einstein A., Infeld L., *The Evolution of Physics: The Growth of Ideas from Early Concepts to Relativity and Quanta*, Simon & Schuster, 1938
2. Isaacson W., *Einstein: His Life and Universe,* Simon & Schuster, 2007
3. Clerk R.W., *Einstein: His Life and Times*, Hodder & Stoughton, 1973
4. Richards et al, *Modern University Physics, Part 2*, Addison Wesley 1973
5. Miller A. *137: Jung, Pauli, and the Pursuit of a Scientific Obsession, Ps.234-235.,* W. W. Norton & Company, 2010'
6. Ash D. *The New Science of the Spirit*, The College of Psychic Studies, 1995
7. Perlmutter S. et al *Discovery of Supernova Explosions...* Lawrence Berkeley National Laboratory, Dec 16, 1997
8. Hawking S. *A Brief History of Time*, Bantam Press, 1988
9. Miller A. *137: P.63.,* W. W. Norton & Co, 2010'

6. The Us of Me - Shane Guffogg

This image is an amazing depiction of interacting vortices setting up electric force fields. - Ash

ESSAY 6

Electromagnetism

"It is simple therefore it is beautiful."

- Richard Feynman

Précis: In the vortex theory the term electromagnetism is applied to the two ways vortex energy can spin. The spin of energy forming the vortex is directed in or out of its centre. That sets up the electric force. This spin is an expression of the innate dynamic nature of the vortex as it extends into infinity. The vortex itself can move, and when it does so it can set up another force known as magnetism. The most common form of magnetism is caused by a rotational spin of a vortex of energy. In physics this is known as its magnetic moment.

To understand the electric force and the magnetic force, think of the vortex in terms of whirlpools and tops. The spin of energy forming the vortex is like a whirlpool. That sets up the electric force. The whirlpool can then rotate like a top. That sets up the magnetic force.

The vortex of energy is more like a ball of wool than a whirlpool. It is spherical rather than conical. As modelled by a ball of string or wool, close to its centre the ball vortex is a three-dimensional spiral, but at a distance out from the centre it is less like a spiral and more like a system growing or shrinking concentric spheres.

The electric force is effective at a distance out from the centre of the vortex, so the electric interaction between two vortices would be more like two lots of concentric spheres banging against each other than two spirals intertwining. A ball of string or wool can either grow of shrink. If you are winding it up it grows. That is the equivalent of a positive charge. If you are knitting and it shrinks that is the equivalent of a negative charge. One doesn't normally think of a ball of wool as a vortex, but if you were to knit at the speed of light it would be a dynamic vortex of wool. If you rotate the ball that is the equivalent of a magnetic moment.

The primary spin in or out of the vortex, which sets up an electric field, involves a change in the intensity of vortex energy. The secondary spin of the vortex, which sets up a magnetic field, does not incur such a change. However, the magnetism can move. If this tertiary activity incurs a change in intensity of vortex energy it can act as an electric field. As an invisible extension of matter, moving magnetism can hammer electrons and generate an electric current. A growing or decaying magnetic field incurs a change in intensity of vortex energy. That is why growing or decaying magnetic fields can induce a flow of electricity. These are the forces of vortex energy that drive our electric civilisation.

~

Essay: In the vortex theory a ball of wool is used as a model to depict the ball vortex of energy because it shows not only a spherical vortex but a system of spin where the direction of spin can be either in or out of the ball. When someone winds a ball of wool it expands. When they are knitting it contracts. The ball of wool is a dynamic vortex when it is either growing or shrinking; especially if this were at the speed of light.

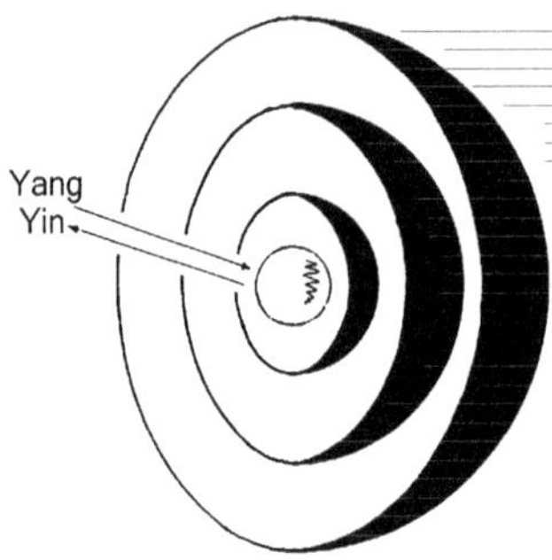

Fig 52 Yin Yang

Imagine you have a ball of wool in your hand and rotate it. That depicts the vortex energy flowing in or out and spinning roundabout. The two forms of spin, in and out or roundabout, set up the two forces associated with a quantum vortex of energy known as electromagnetism. The primary in or out spin forming a subatomic particle, such as an electron, sets up its electric charge. The secondary top-like spin sets up its magnetism. This is called its magnetic moment.

The Chinese would designate the opposite directions of spin in ball vortices of energy Yin or Yang. In the classical description, Yin stands for contraction and Yang stands for expansion. In the ball of wool analogy, Yin would represent the ball shrinking when someone is knitting. Yang would represent the ball growing when someone is winding wool into a ball. If the opposite directions of spin represent negative or positive

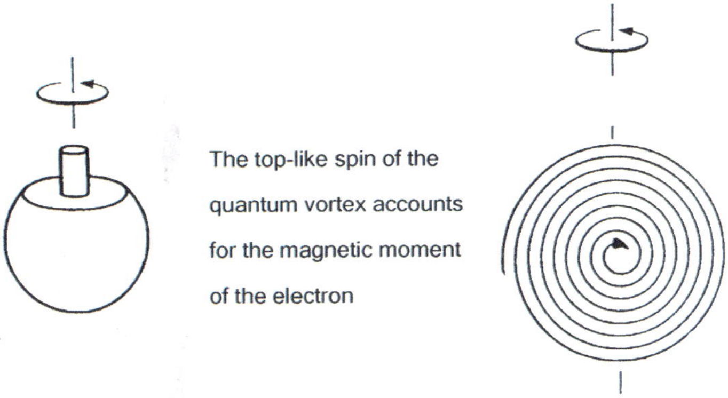

Fig 53 The spin of an electron sets up its natural magnetism

charges, Yin would represent a negative charge where energy is flowing into the centre of a vortex and Yang a positive charge where energy is flowing out of the centre of a vortex.

In the vortex theory the structure of the electron is a spherical vortex of energy but in its action it behaves as a system of growing or decaying concentric spheres. This is because at a distance from the centre of spin the energy in the spherical vortex tends to flow in or out as nested spheres. The ball of wool models the way that with displacement from its centre the ball vortex appears to be less like a three-dimensional spiral and more like a system of concentric spheres. This is apparent in its behaviour.

At a distance from its centre the ball vortex acts as a system of growing or shrinking concentric spheres of energy. This is apparent when positive or negative charges overlap. They behave not as interwinding spirals but as growing or shrinking spheres that come up against each other and either push apart or pull together depending on their relative directions of three dimensional spin; whether the spheres are contracting or expanding. These effects are illustrated in a simple diagram that depicts overlapping concentric spheres of energy.

A pair of ball vortices of energy can be represented as adjacent concentric circles with arrow lines depicting whether the concentric spheres are growing or shrinking. If the circles are overlapped and arrow lines are drawn at their points of contact, lines of force emerge that are characteristic of electric fields. Overlapping circles representing positive charges show opposing action to cause repulsion. Overlapping circles representing negative charges also show opposing action to cause repulsion. However, circles representing a positive charge overlapped with circles representing a negative charge show an action

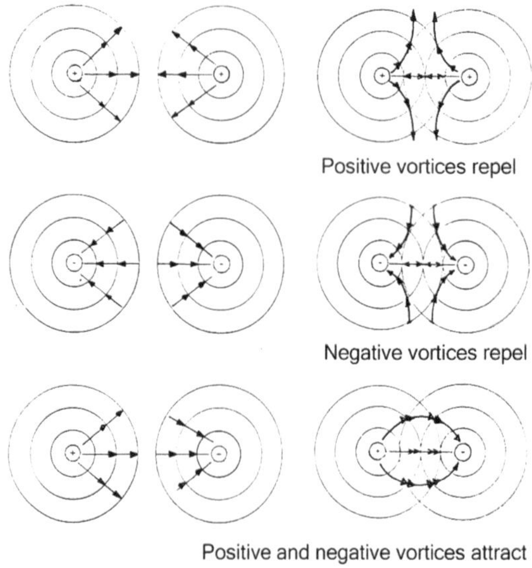

Fig 54 The vortex theory for electric interactions

that would cause attraction. These diagrams show clearly and simply why like charges repel and unlike charges attract, and they match the textbook diagrams of lines of electric force.

The French scientist, Charles du Fey was the first to recognise that like charges repel and unlike charges attract. If when they meet one ball vortex is expanding and another is contracting then their energies would be flowing relative to each other in the same direction. Their relative interaction would be unidirectional. They would then act with each other and so would attract. If both ball vortices are expanding or both are contracting then their energies would be flowing relatively in opposite directions where they meet. Their relative interaction would be antidirectional. They would then act against each other and so would repel. You could imagine them butting up against each other aggressively when their energy is antidirectional to cause repulsion and flowing with each other in harmony when their energy is unidirectional to cause attraction.

Imagine interacting vortices of energy as swans. They flap angrily when they come up against each other in opposing directions but glide gracefully together when they are in the same direction.

The arrow lines indicating the directions of energy movement in a ball vortex

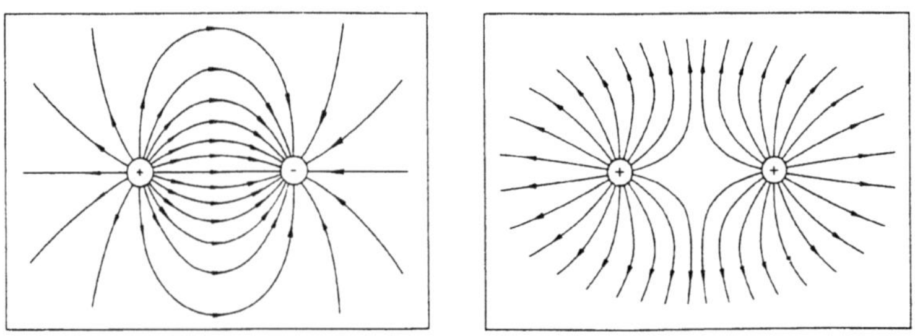

Fig 55 Textbook diagram of lines of electric force

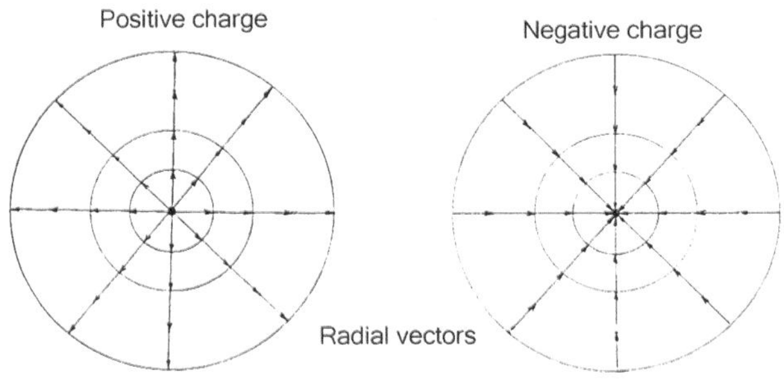

Fig 56 Radial vortex vectors

are 'vectors'. A vector is the direction in which a motion is effective. The vector lines for the concentric circles are along their radii. This shows that electric charges, behaving as growing or shrinking spheres of energy, interact effectively along their radii. Electric charge interactions can be described in terms of the radial effects of vortex energy. I describe them as radial vortex vectors.

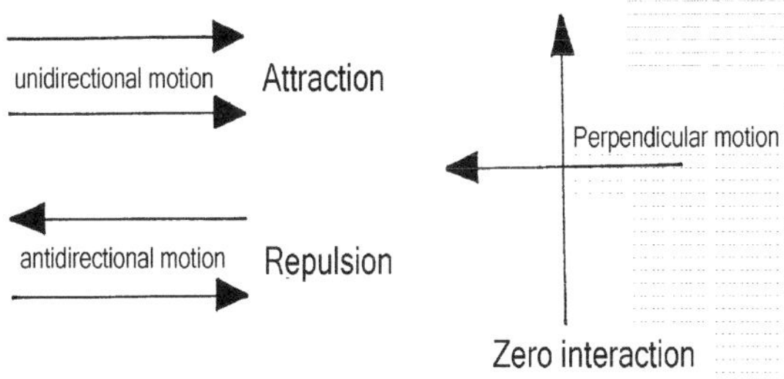

Fig 57 Laws of vortex interactions

Mid-way between attraction and repulsion is zero interaction. Mid-way between movement in the same direction and movement in opposite direction is movement crossing at right angles.

Growing or shrinking concentric spheres of vortex energy forming an electric charge is, in effect, motion along their radii. By contrast the rotational spin of the spheres, causing magnetism, would be effective at a tangent to them. Because a tangent is perpendicular to the radius it subtends, the rotation of vortex energy causing magnetism would be effective at right angles to the flow of vortex energy causing its charge. Generally electric fields occur at rightangles to magnetic fields. The vortex theory show why this is so.

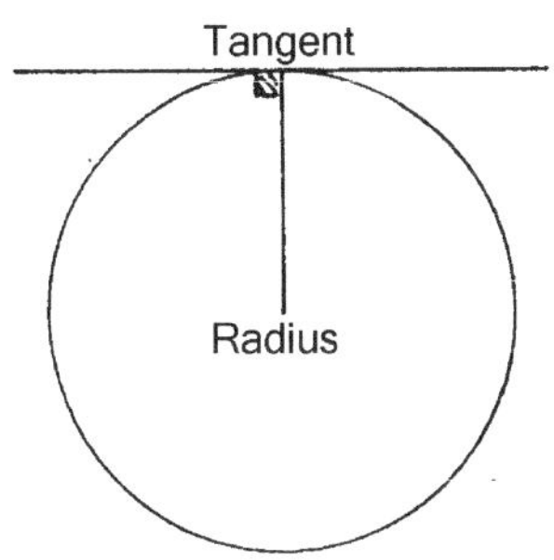

Fig 58 Tangent and Radius

Because the in or out spin of vortex energy forming an electron is at right angles to the rotational spin that sets up its magnetic moment, these two distinctly different types of spin – one in and out and the other round and about - would set up distinct fields of force. That is why magnetism is distinct from electric charge. As the two types of spin are effective at right angles they would not interact. Sure enough, a stationary magnetic field does not interact with an electric field and vice versa.

However, a moving magnetic field can interact like an electric field. This is because a new motion has been laid onto the rotating vortex of energy. We are dealing with three motions of energy here. The first is in or out, the second is round and about and the third is back and forth. As the back and forth motion can head-on to a vortex particle, it would be in the same direction as the flow of energy in the vortex, up or down it radius, that sets up its electric charge. The movement of the magnetic field can therefore be effective as an electric field and induce a flow of electricity. You could imagine the moving magnetic field as an invisible extension of matter, hammering the electrons. This is how a dynamo works. The magnetism extending from spinning magnets in the dynamo act like hammers on electrons. The hammering induces shock waves that pass from electron to electron in the copper of a wire. That is how electricity flows down the wire. Imagine electrons in a wire as railway carriages in a station. The moving magnetism acts like a shunt. The carriages hardly move but the shunt sends a shock wave down the row. That models electricity.

The 3D spin of energy in or out of a vortex setting up an electric force involves a change in intensity of vortex energy, whereas the rotational roundabout spin setting up a magnetic force is characterised by a uniform intensity of vortex energy. This is because the growing or shrinking spheres of energy will be changing in intensity, whereas their rotational spin at a constant speed involves maintains a uniform intensity of energy. It follows that a change in intensity of vortex energy can act as an electric force whereas a uniform intensity of energy maintains a static magnetic force.

A growing or decaying magnetic field involves a change in intensity of vortex energy, which enables it to act as an electric field. That is why growing or decaying magnetic fields can induce electric currents. The changes in intensity of vortex energy in a magnetic field, growing and decaying fifty times a second, is the source of the alternating current of electricity that drives our civilisation. You can feel the vibration of these changes as a slight buzz when you gently stroke an appliance plugged into an AC electric circuit. The buzz is not harmful.

In the atom, electrons occur in pairs. Each member of the pair spins in opposite direction so their magnetic moments cancel each other out. This is why most atoms are non-magnetic. Natural magnetism arises from atoms which contain an unpaired electron. Because this magnetism is not cancelled it confers magnetism on the atom as a whole. Atoms of iron are an example of this. However a lump of iron only becomes magnetic when its atoms are all aligned in the same direction. This can be achieved by stroking the lump with a magnet. If the lump is heated the atoms lose their alignment and the lump of iron can lose its magnetism.

If nothing exists but the motion called energy, all forms and forces in the Universe would consist of forms of motion interacting with each other and some of the interactions could be quite complex as one form of motion superimposes or layers on another. The radial flow of concentric spheres of energy is the primary motion in the vortex giving rise to electric charge. The vortex can then rotate, which gives rise to its natural magnetism. The rotating quantum vortex can then vibrate or move. These tertiary motions are layered upon the primary and secondary motions of a quantum vortex. They may contribute to or cancel out the primary or secondary force fields. All of this depends on the relative directions of the overlapping and interacting energy. The directions of these layers of motion set up the layered vortex vectors, which account for electromagnetism, and the way these vortex vectors set up the interactions of electromagnetism follow self-evident rules.

The interaction rules between vortex vectors are simple and self-evident. Remember the swans. When vortex vectors move with each other in the same direction they attract. When they move against each other in the opposite direction they repel.

To appreciate the interactions of electromagnetism it is necessary to take into account that if layered vortex vectors move relative to one another along the radii of the vortices they behave as an electric field. If their relative movements are at a tangent to the vortex, they behave as a magnetic field.

The vortex rules of interaction apply to both electric charge and magnetism. For example there is a magnetic repulsion between two electrons spinning in same direction. This occurs because the rotating vortex vectors are in the opposite direction where they meet.

There is a magnetic attraction between two electrons spinning in

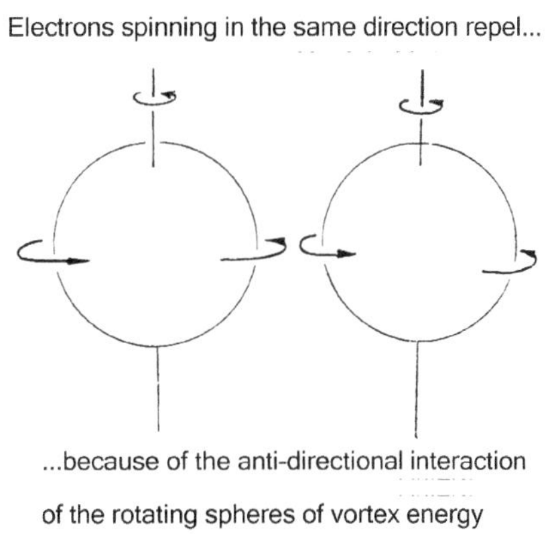

Fig 59 Spinning electrons that repel

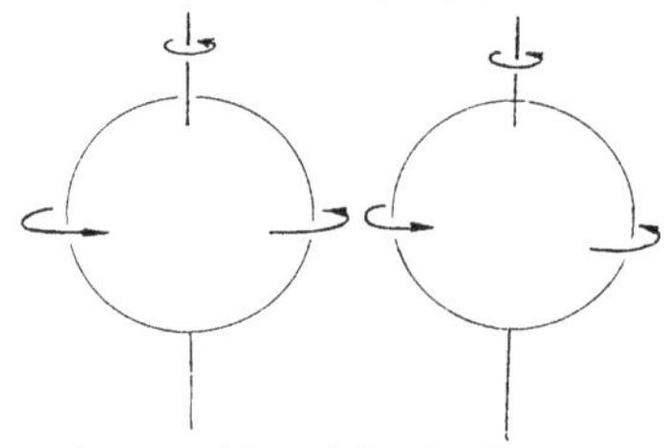

Fig 60 Spinning electrons that attract

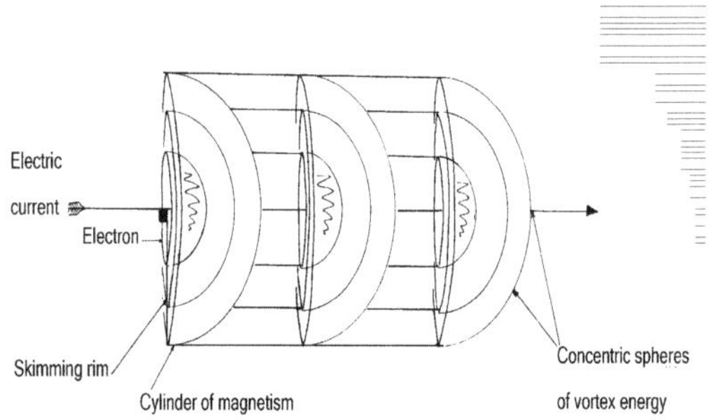

Fig 61 Magnetism around an electric current

opposite direction. This is because the rotating vortex vectors are in the same direction where they meet.

When electron vortices flow in an electric current, vortex vectors associated with this motion are effective at a tangent to them. These set up concentric rims of magnetism that surround each moving vortex at right angles to the current of electricity. The concentric rims of magnetism add up to form concentric cylinders of magnetism. Imagine slicing an onion then skewering it through its centre. In the line of the skewer, representing the current, the moving onion rings would form concentric cylinders of onion representing the magnetic field.

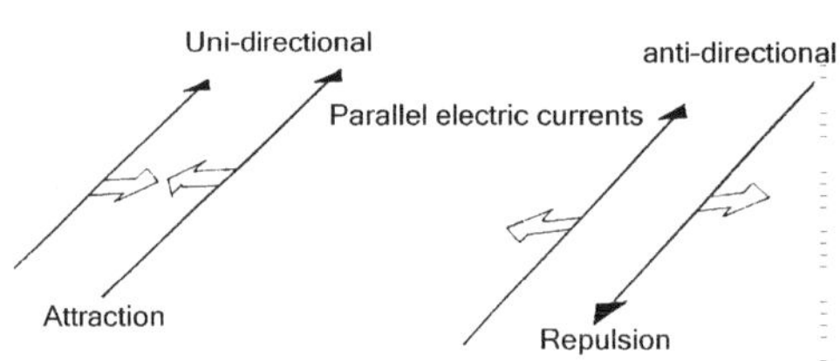

Fig 62 Ampere effect

If two current-carrying conductors are laid side by side they will attract if the current is flowing in the same direction but repel if it is flowing in the opposite direction. This effect was first observed by the French scientist André Ampere.

When the concentric spheres of vortex energy extending from the two parallel electric currents move against each other in opposite direction, they repel. When they move with each other in the same direction, they attract. The vectors are effective at a tangent to the concentric spheres of vortex energy, so these interactions between the current-carrying conductors are magnetic.

In the 1830s, in the basement of the Royal Institution of Great Britain, a young man called Michael Faraday was experimenting with magnets, wires, and electric currents generated by batteries. He found that a stationary magnetic

Fig 63 Michael Faraday

field did not interact with a stationary electric field. But Faraday found that a moving magnetic field interacts with an electric field. He discovered that a moving magnet could induce a flow of electricity in a wire. He used this discovery to invent a rudimentary dynamo which paved the way for modern electricity generators.

Each photon consists of two lines of the movement of light travelling in transverse waves at right angles. Because perpendicular fields of energy do not interact they can coexist in the same quantum bundle without losing energy to interaction. However, in the nineteenth century, because of this perpendicular arrangement of the fields of energy in light and radio waves, it was assumed that they were electric and magnetic fields radiating through space. It was only natural that the term electromagnetism was applied to the propagating waves of energy because they could interact with electric circuits in radios and suchlike. But magnetic and electric fields are both effective in the world of matter, whereas in the twentieth century it was discovered that only one of the fields in the photon reacts with matter. The other field is presumptive antimatter and plays no part in the world of matter. This means that the electromagnetic theory for light no longer holds true and the term electromagnetism should not be applied to radiant energy.

Experiments with antimatter confirm the importance of vortex vectors in the quantum vortex of energy. The primary direction of energy in the quantum vortex appears to be paramount. If the direction of the of energy forming a subatomic vortex – causing its electric charge – is reversed, then the effect of the secondary motion setting up its magnetism is also reversed. Particles with the same sign of charge, but spinning in opposite directions, experience a magnetic attraction. If the sign of charge of one of them is reversed, then the magnetism will reverse and set up a force of repulsion. This effect has been observed between particles of matter and antimatter. Just prior to annihilation, electrons and positrons undertake a 'death-dance' called positronium. If the electron and positron are rotating in the same direction they attract each other and consequently annihilate sooner. If electron and positron rotate in the opposite direction the magnetic repulsion delays their annihilation.

Opposite charges cancel out each other's effects. Space is electrically neutral because there are an equal number of opposite charges in existence. Atoms are electrically neutral because they contain an equal number of opposite charges. Why should neutrons be different? They are clearly composed of oppositely charged particles. Why would they be different from atoms?. Why would the neutrality of the neutron defy a fundamental law in physics, if not to protect Heisenberg's uncertainty principle? How long is science to be compromised for the sake of a theory of uncertainty that has introduced craziness and maddening complexity into physics? It is only a matter of time before the truth will out. The danger is that when the uncertainty principle goes, as surely it will, it could take quantum mechanics with it. The imperative is to save quantum mechanics. It could be protected from the inevitable collapse of the uncertainty principle if this underpin were to be replaced with a more credible foundation. Quantum mechanics is the Claridges of physics. Like the work under Claridges Hotel in London, the necessary changes to save the structure have to be made without disturbing the parties on the floors above. Let the physicists party with whatever models taste best in their cocktails of equations but, for those who might prefer a sip of simplicity in the cool of common sense, two props of Newtonian normality are ready to be introduced into the basement of quantum mechanics. Two quantum laws of motion, based on Newton's first law of motion, have the potential to be pillars of quantum mechanics in place of the uncertainty principle. If nothing exists but the motion called energy it makes sense that the most fundamental laws in the Universe would be laws of motion.

7. The Now of Then - Shane Guffogg

In this image I see cylindrical magnetic interactions between current-carrying conductors - Ash

8. Rhyme of Reason - Shane Guffogg

I see interacting waves and vortices in this beautiful image - Ash

ESSAY 7

The Quantum Laws of Motion

"Quantum mechanics, you never understand it, you just get used to it."

- Richard Feynman

Précis: In standard physics it is assumed that the existential basis of energy is the wave. In the vortex theory energy is thought to occur in the form of vortices as well as waves. The vortex form of energy is believed to be the basis of matter and the wave form of energy is the basis of light. The word 'quantum' means a particle of energy, so in the vortex theory the word quantum is applied to the vortex as well as the wave.

Quantum particles of energy interact. Vortex on vortex interactions cause the forces of electric charge, magnetism and gravity. Wave on vortex interactions are considered to be the cause of wave-particle duality, kinetic energy, nuclear energy and the transitions between mass and energy.

In the unified field of the vortex theory an electron or a proton, as a quantum vortex of energy, would be a three-dimensional spiral space path to a propagating quantum of waves. The quantum vortex is passive whereas the wavy quantum is active. The wave form quantum would drive into the vortex as if it were a car on a spiral road. Driving into the vortex it would be captured. In the vortex theory the quantum vortices trap the quantum waves of energy.

The electron vortex has very little mass so it would move under the impact of a quantum of waves. Only a part of the propagating wavy quantum would be caught. The rest of it would stick out like a tail and, being a wave train, it would force the electron vortex forward in wave motion. The electron with partially captured wave-form energy would thus become a wave-particle. This is how wave-particle duality is explained in the vortex theory. It also provides an account for kinetic energy.

The two forms of energy can transition between vortex and wave. These transitions are governed by two quantum laws of motion. The first law is that waves and vortices are stable forms of motion that change only if they are forced to. The second law is that if the force of change ceases they revert back to their original forms of motion. These quantum laws of motion are derived from Newton's first law of motion. In the vortex theory quantum mechanics is derived from Newtonian mechanics.

The first quantum law of motion concerns energy capture. Imagine the wave form of energy undergoing entanglement and becoming captured much as convicts would enter prison. The prison takes no action in the entanglement and becoming captured. The captives walk into the prison which then holds them behind its bars. The second quantum law of motion concerns escape from capture. Imagine wavy quanta escaping capture as prisoners escaping prison. On escape the convicts ran away and immediately revert to freedom. Likewise escaping captured energy would revert immediately to its original state of wave propagation.

The proton vortex is much more massive than an electron vortex so it can withstand the impact of a quantum of waves. The quantum of wave energy drives right into it and becomes completely captured. Vortex motion is then imposed or enforced on the wavy quantum. In the quadrillions of years of their existence proton vortices become completely saturated with energy they have captured. If two protons then collide with a force sufficient to overcome their electric force of repulsion – in the Sun or a hydrogen bomb – they converge, and a small amount of the captured energy is squeezed out. It immediately reverts from enforced vortex motion to its natural state of wave motion. That is nuclear energy. Meanwhile, the remaining captured energy swirls between the converged proton vortices and binds them together. That is the nuclear binding force.

In line with the first quantum law of motion, in high-energy research facilities the wave form of energy is transformed into the vortex form when it is forced through vortices of energy in high-energy collisions. One could imagine researchers in a particle accelerator as chefs forcing batter through a doughnut-making machine. The batter comes out in the shape of a doughnut but it doesn't last long if it falls to the floor. It immediately reverts to a mess of batter. Likewise the energy forced through nuclear vortices in high-energy collisions comes out as swirls of vortex motion, which don't last long. In line with the second quantum law of motion they immediately revert to their original wave form and radiate away. This is why the particles generated in high-energy research are very short lived.

Fig 64 Albert Einstein

Essay: In *The Evolution of Physics* published in 1938 [1] Einstein wrote:

> "Belief in an objective reality has led to great scientific advances throughout the ages. Without the belief that it is possible to grasp reality with our theoretical constructions, without belief in the inner harmony of our world, there could be no science. This belief is and always will remain the fundamental motive for all scientific creation."

Einstein had two major contentions with the architects of quantum mechanics. One was the uncertainty principle and another was the discredit of realism. Almost immediately, Einstein realised these were catastrophes for science but despite his authority he was overruled. A third major casualty that resulted from the chaos of the quantum revolution was the loss of distinction between matter and light at the quantum level of reality. Always before it was taken for granted that matter was fundamentally different from light and the invention of the vortex atom theory was an attempt to explain this. However, when the theoretical models of the nineteenth century were dismissed as naïve realism, the distinction between matter as vortex motion and light as wave motion was lost.

The major lights in late nineteenth century physics, including Lord Kelvin, James Clerk Maxwell and Sir J.J. Thomson, never questioned the distinction between matter and light. Up until 1910, students at Cambridge were taught that light was based on wave motion and matter was based on vortex motion. When the vortex atom theory was abandoned, its implication that there might be two fundamental forms of energy was lost, then the quantum distinction between matter and light was lost completed by another discovery.

While Einstein was establishing that waves of light were particles, the French Count Louis de Broglie was proposing that particles of matter could be treated as waves. When in 1927, two Americans, Clinton Davisson and Lester Germer, demonstrated that a beam of electrons could produce wave patterns, Louis de Broglie's theory of wave-particle duality was accepted and he subsequently received the 1929 Nobel Prize in physics. Erwin Schrödinger then developed wave equations which compounded the merge of matter and light and, with the established frequency difference between the quantum states in Bohr's atom, it is hardly surprising that physicists considered they could account for matter as well as light in terms of wave function. From then on it was difficult to visualise any distinction between matter and light at the quantum level.

However, the visualisation of quantum spin provided by the new vortex theory suggests a clear distinction between matter based on vortex motion and light based on wave motion. Physicists may be concerned that the reintroduction of the vortex model is a return to naïve realism but the visualisation of quantum spin it is not naïve, it justifies realism and satisfies Einstein's concern that the quantum world should be knowable. The knowability of quantum reality was established thousands of years ago in ancient India in a practice of yogic science based on meditation. The vortex spin underlying matter was perceived directly by yogis using powers attained through meditation and the use of these powers to probe the smallest realms of matter were recorded as early as 400 BCE in the Yoga Sutras of Patanjali, where the results of meditation were described in detail. In Aphorism 3.46 it states that through meditation the yogi gained an extended faculty of observation from the practice of the anima siddhi - an ability developed by yogis to shrink their consciousness commensurate with the very small.

If subatomic particles of matter are based on vortex motion and photons of light are based on wave motion then many of the structures and actions in the Universe could be a consequence of their interactions. The forces of electric charge, magnetism and gravity have been shown to be a possible consequence of vortex on vortex interactions but the structure of the atom as envisaged by Neils Bohr, and the phenomenon of wave-particle duality predicted by Louis de Broglie as well as other phenomena, such as thermodynamics, kinetics, and the outcomes of high-energy research could be considered in terms of wave on vortex interactions.

The interactions between waves and vortices of energy are based on the premise that space and mass are the same phenomenon; vortex energy. In the unified field of vortex energy the approach to the centre of mass in an electron or a proton is treated as an increasingly tight three-dimensional spiral space path. The quantum vortex is passive whereas the quantum wave is active. On encountering a subatomic particle a wave-train quantum or photon would find itself in spiral space. Following the increasingly curved path of space it would be the equivalent of a car on a spiral road. It would drive into the vortex because it would be in an increasingly tight spiral of space. The wave quantum would be trapped by the quantum vortex because unlike the car in my analogy it would have no reverse gear. That is because light always flows forward. It never goes in reverse. For the wave-train quantum or photon there would be no way out. It would become entangled and captured by the vortex.

Vortices are energy traps not by action but by shape. This is illustrated by the car and road analogy. The action of getting stuck is taken by the car. The road is entirely passive in the process. I have a vivid recall as a child of an outing with my eccentric father in his old Rolls Royce. He had taken a wrong turn on a narrow Cornish lane and we ended up in a farm yard confronting a steaming heap of dung. While our Pa insisted it was the lane that caused us to get stuck me and my brothers and sister were in fits of laughter because we knew it was entirely the fault of his driving that we ended up in that smelly situation.

I explained the wave-particle duality of an electron by suggesting a propagating string of energy in a photon, effective in the world of matter, would follow the spiral path of space, presented by an electron vortex, and drive into its centre. Meanwhile, the other undulating line of light in the quantum of wave energy, effective in the world of antimatter, would follow the mirror spiral of space and drive itself into the centre of the positron vortex. At the singularity centre of the vortex, shared by the twin electron-positron pair, the two lines of energy would butt up against each other, preventing further progress of either. The photon would thereby be captured irretrievably by the subatomic matter-antimatter pair. I called this capture theory. The vibrating strings of energy in the photon could not stop moving, however, because pure movement is their nature. The trapped photon would therefore impose its wave motion onto the positron and electron pair and drive them forward simultaneously into wave motion. At the same time the energy in the photon, caught in the twin quantum vortices, would fill their spiral space and so increase their mass. The result of this interaction would be a mutual sharing of the wave and vortex forms of motion. The vortex would take on the wave form of motion and the photon would partially assume the vortex form of motion. I called this partial capture. The result of the wave-vortex quantum entanglement is a wave-particle.

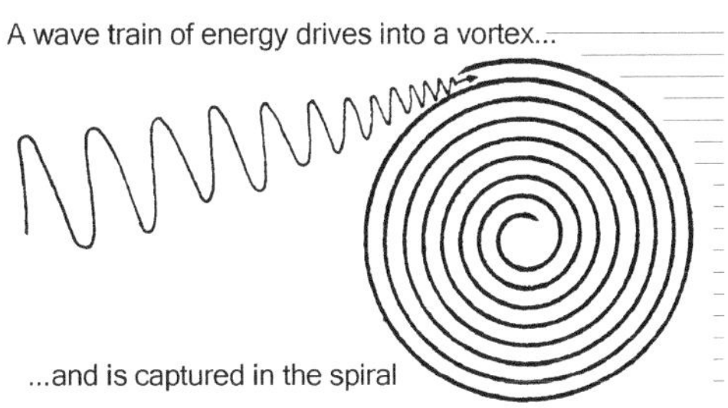

Fig 65 Capture theory

Fig 66 The photon

Imagine a photon partially captured by an electron as a vortex with a wave train of energy sticking out of it driving it forward in wave motion. This wave-particle duality of electrons is illustrated by tadpoles. A tadpole swims in a wavy line because it is driven by a wavy tail. In much the same way, a vortex particle of matter, such as an electron, would be driven into constant wave motion by a partially captured quantum of wave energy attached to it like a tail. The reason for this situation could be that the electron has insufficient inertia to withstand the impact of a quantum of wave energy. With the electron always moving ahead of the wave quantum it could never enter the electron completely.

Fig 67 The tadpole analogy

The concept of inertia originates from Sir Isaac Newton's first law of motion

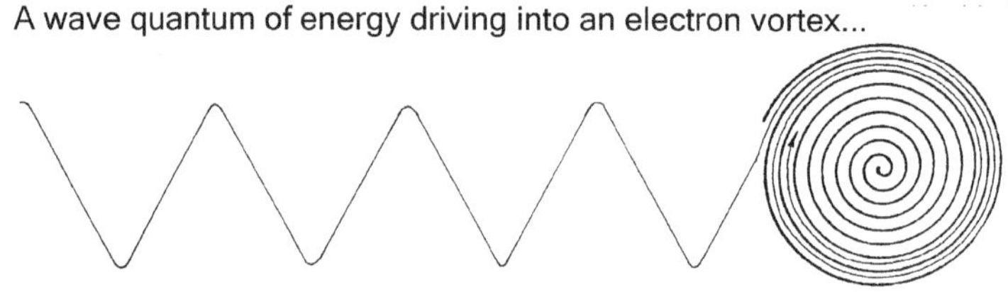

Fig 68 Partial capture

that a body will to stay at rest or keep moving unless a force acts on it to do otherwise. It struck me that if the Universe is formed of two fundamental forms of motion, the wave and the vortex, then the existence of the Universe must depend on their stability That suggested to me that Newton's law of things not changing unless they are forced to change, had to be one of the most important laws in the Universe, because it was the fundamental law of stability. It made sense to me that Newton's first law of motion should apply at the quantum level so I suggested that a quantum of energy would maintain its form or state of motion unless change is forced upon it, and if the force is removed it would revert to its original state. Based on that premise I formulated two self-evident quantum laws of motion to govern the interactions of vortices and

Fig 69 Isaac Newton

waves of energy as the basis for a new approach to quantum mechanics based on Newton instead of Heisenberg.

The Quantum Laws of Motion

1. A quantum of energy will maintain its form of motion unless a change of form is imposed upon it.
2. If the imposition ceases the quantum of energy will revert to its original form of motion.

A spring illustrates the quantum laws of motion. A spring will only stretch or compress if a force is applied to it. If the force is removed the spring will immediately revert to the state it was in before the force was applied. This is the inertia of the spring.

The quantum laws of motion are based on the assumption that every particle of energy has a predetermined form of motion, which is either a train of waves or a vortex. The form of a particle of energy can be changed but as soon as the influence of change is removed it will spring back to its original form. If particles of energy are more like thoughts than things, the predetermined forms of a quantum of energy could be imagined as a memory.

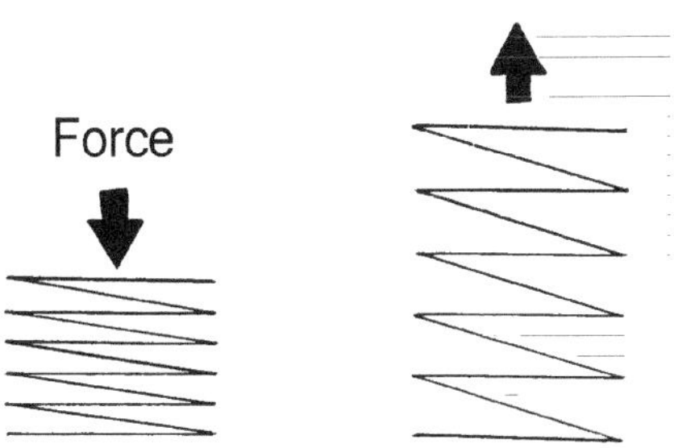

Fig 70 The spring analogy

The vortex sets up static inertia so the inertia of the quantum vortex is static. Its inertia is to stay put unless it is moved. The inertia of the quantum of waves is kinetic. Its inertia is to keep moving unless it is stopped. For this reason the quantum laws of motion could be described as the quantum laws of inertia. Many events and much activity in the Universe is a consequence of the endless tussle between these opposite inertias.

To understand the opposition of inertias in quantum reality imagine a monkey and a jar of nuts. The monkey represents a photon or a quantum of waves, and the jar of nuts represents a quantum vortex. The inertia of the jar of nuts is to remain on the ground unless it is moved. The inertia of the monkey is to keep moving unless stopped. The monkey is attracted to the jar of nuts and

Fig 71 The monkey analogy

puts its hand in the jar to grab some nuts but when its hand is full of nuts it can't pull out. Unwilling to let go of the nuts the monkey is captured by the jar. The jar takes on the motion of the monkey which is running round with the jar of nuts swinging at its wrist. That illustrates a wavy quantum of energy driving into a subatomic vortex and imposing wave motion on it. The jar becomes heavier for the monkey's hand. That illustrates the quantum vortex becoming more massive for the input of additional energy into its spiral structure. Each system of energy imposes its form of motion on the other. This is the first quantum law of motion. If the monkey lets go of the nuts and pulls its hand out of the jar it will be free to run away, and the jar of nuts will end up on the ground much as it was before the monkey came along. That illustrates the second quantum law of motion.

In the state of partial capture the quantum wave particle and quantum vortex particle become a composite of two particles of energy. Each acts in opposition to the other so their coming together can be described in terms of opposing inertias. The kinetic inertia of the wave form of energy pushes the vortex form forward into wave motion, and the static inertia of the ball vortex resists the attempt to be pushed forward. Balance is found between these opposite inertias. In this balance the opposite inertias are shared. As the vortex takes on the kinetic inertia of the wave-quantum and the wave-quantum adds to the static inertia of the vortex, a bonded state is set up between them. It is this balance of opposing, shared inertias that sets up wave-particle duality.

The quantum laws of motion provide a way of accounting for wave-particle duality that opens the door to a new version of quantum mechanics. This is needed if a new underpin is to be found for quantum mechanics to replace the uncertainty principle. The quantum laws of motion could lead to a new understandable version of quantum mechanics that is compatible with Newtonian mechanics. On this count alone a new quantum mechanics based on the quantum laws of motion has an advantage over the quantum mechanics based on Heisenberg's principle. Common sense speaks in favour of the quantum laws of motion as they work well in explaining many experiments and discoveries in physics.

The Egyptian teacher Hermes Trismegistus, who lived three generations after Moses, is credited with establishing the Greek civilisation and the foundation of science and medicine. Medicine still bears his symbol of the caduceus.

Ernest Lord Rutherford recognised the original role of Hermes in physics by incorporating him in his coat of arms. Hermes anticipated quantum physics.

Fig 72 Ash at the Rutherford Memorial, Nelson N.Z.

This anticipation is in his name. 'Hermes' means motion, fluidity and activity in ancient Greek. The nearest equivalent in modern English is 'energy'. He was named Hermes in Greek and Mercury in Latin because he was so emphatic that the dynamic state is fundamental to everything in the Universe, which he treated as a mind. Hermes is famous for the principle "as above, so below". In science this is called symmetry. An example of ..."as above, so below" symmetry is the symmetry between Newton's first law of motion and the quantum laws of motion. Another principle proposed by Hermes was that the gender of masculine and feminine applies at every level in the Universe.

In many ancient societies, matter was considered to be feminine and light was considered to be masculine. In Latin the word matter comes from Mater meaning mother. The ball vortex fits with the feminine principle as it is spherical and receptive like an ovum. The wave-kinetic quantum fits with the masculine principle as it appears like an undulating sperm. The quantum vortex is the feminine form of energy. The wave quantum is the masculine form of energy.

Just as the sperm drives into the ovum to initiate new life so the wave quantum drives into the ball vortex to initiate the kinetic activity that underlies life. The man penetrates the woman, the sperm penetrates the ovum, the wave quantum penetrates the quantum vortex. As above so below... at every level in the Universe there are similar gender patterns.

The first quantum of energy that drives into an electron vortex penetrates the innermost and tightest spiral space path in the electron vortex. This sets up a stable bond between a wave quantum and an electron vortex which is exemplified by marriage. After marriage it's, 'out to work, you lazy hellion!' In physics, the useful consequences of interactions between heat, light and matter are called work.

The strong wave-vortex bond between the initial quantum, akin to a husband,

and an electron vortex, akin to a wife, accounts for the stable ground state electron in an atom. But just as wives can be attractive to other men, so the ground state electron vortex is attractive to other quanta of energy. They too can enter its spiral space. With its inner tightest spiral space already occupied by the first quantum the next quantum can only penetrate the looser more peripheral regions of the three dimensional spiral space. They cannot bond as strongly with the electron vortex. Just as the social bond between a lover and a wife is weaker than between a husband and a wife, so the bond between the electron vortex and the second quantum is weaker than with the first.

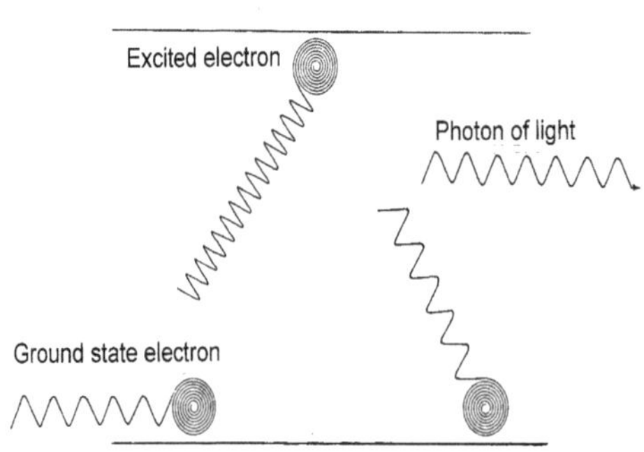

Fig 73 The quantum leap

With the arrival of a lover the wife becomes excited. Likewise, when electrons in atoms absorb additional energy they become excited and leap from the ground state orbit into an excited state orbit. In quantum physics this is called the quantum leap.

Because of the weaker bond the additional energy captured by an electron vortex can easily break free. This is exemplified by the weakness of le trio. Just as a disillusioned lover might break away and leave the wife to return to monogamy, so in an atom the weaker bonded quantum can break free enabling it to fly from capture as a photon of light. The electron vortex then drops from the excited state back to the ground state and a photon of light is emitted. This has a frequency – represented by colour – corresponding to the energy difference between the ground state and the excited state. This is how the vortex theory for subatomic particles accounts for the colour spectra of atoms, which was one of the discoveries that defeated the theory of the vortex atom.

If you burn a candle, the heat released by the combustion will excite electrons in carbon atoms floating in the hot air currents rising from the burn. Excited electrons leap to a higher orbit then drop back to the ground state again as they release quanta of energy as photons of light. These appear as a flame, and the frequency of the yellow light of the photons coming from the flame represents the difference in frequency between

Fig 74 The candle flame

the ground state and excited state in the carbon atoms that absorbed energy from the combustion of the wax. There are parallels in the old flame leaving le trio and the flame in physics.

The transition of electron vortices between the ground state, excited state, and ground state again, with the release of photons, is the basis of electric illuminations including the incandescent light bulb and LED lights as well as the colour spectra of atoms. The vortex theory could have accommodated the spectral lines and colours of heated atoms had the Victorian physicists been aware of subatomic particles; but perhaps the subject of subatomic sex might have met with resistance from the prudish Victorians.

If the chemistry is right wives in an excited state sometimes leave their old homes and settle in new ones. As above, so below... so it is in atoms. Electrons in the excited state can leave one atom and settle in another. This is the basis of chemistry. In the chemistry lab, Bunsen burners are used to initiate a chemical reaction because heat is needed to excite electrons in atoms so they can leap into the excited state then move across from one atom into another. Sometimes when a wandering wife moves into a new home she may nip back, from time to time to the old home. So it is in atoms. Very often electrons will run back and forth between the old atom and the new one setting up a bond between them. In the burn of the candle some electrons leap back and forth between the atoms of carbon from the wax and oxygen in the air, and this binds them together into carbon dioxide. In chemistry that is called covalent bonding.

A wife may go off with her lover into a new home and never return to the old one. Breaking up a home can be an explosive affair that leaves a lot of charge between the houses. This is common in the quantum world too. If a piece of sodium metal is left in chlorine gas, even at room temperature, an explosion is liable to occur as heat is generated when electrons with the hots leap from the sodium atom to the chlorine. The sodium ends up as a positively charged ion, having lost an electron, and the chlorine becomes a negatively charged ion for having gained the electron. After the excitement is over they settle down as common salt in a relationship between the atomic homes based on charge. In chemistry this is called ionic bonding.

In chemistry the vital importance of the wave-particle duality of electrons is apparent. This is a consequence of their low static inertia and inability to withstand the kinetic impact of a wave quantum. It is not easy for a quantum of waves to drive into an electron vortex when the electron keeps moving ahead of it. It has to have a lot of energy to do so. I liken this to driving a nail into a swinging gate. It's not easy because the gate moves with the hammer blows. The blows have to be very energetic for the nail to secure a hold and, even then, most of the nail would be left sticking out of the swinging gate. Only part of the nail would be held by the gate. This analogy models a high-energy wave quantum partially captured by a light electron vortex.

Hammering a nail into a gate post is much easier than hammering it into a swinging gate. This is because the inertia of the post enables it to withstand the hammer blows. With a few easy strikes the nail sinks in. So it is with the proton vortex; with a mass equivalent of 1,836 electrons it has sufficient static inertia to withstand the impact of a wave quantum of energy driving into it. Just as old gate posts are slightly more massive for the nails driven into them so proton vortices would be more massive for the wave-kinetic quanta they have captured over the quadrillions of years of their existence. A nail driven completely into the post represents the complete capture of a wave-quantum. In line with the first quantum law of motion the completely captured energy swirling in the three-dimensional spiral space within the proton vortex would contribute to its mass. The diameter of the proton could represent the extent of its ability to hold energy in complete capture in its spiral space path.

I use a crab pot analogy to illustrate the complete capture of energy entangled in protons. Crabs crawl into sedentary pots and are caught by virtue of the shape of the pot. Crab pots do not act in the process of crab capture. In like manner, the wave train quantum of energy would drive into a vortex of energy and be captured by the spiral shape of space in the vortex. The vortex of energy presents a three-dimensional spiral space path to wave energy. With its static inertia the quantum vortex remains passive. It is not active in the process of entanglement and capture.

Fig 75 The crab pot analogy

In the crab pot analogy, just as the captured crab adds weight to the pot, so captured energy contributes to the mass of the capturing vortex. If the crab were to escape it would be free to crawl away and the pot would be lighter for its loss. In line with the second quantum law of motion, if captured energy escapes the quantum vortex its mass would be transformed into energy and the quantum vortex would be less massive.

In the high temperatures of the Sun, or an exploding thermonuclear bomb, protons have sufficient kinetic energy to become wave-particles with very high velocities that can collide with an immense force sufficient to overcome their charge repulsion. As they do so they converge. When they converge their capacity to contain captured energy is reduced and a small amount is squeezed out to be evicted as wave form quanta of energy. In line with the second quantum law of motion this energy is lost from capture as radiant energy. That is nuclear energy. This simple account for nuclear energy is a milestone in physics because Richard Feynman said of nuclear energy:

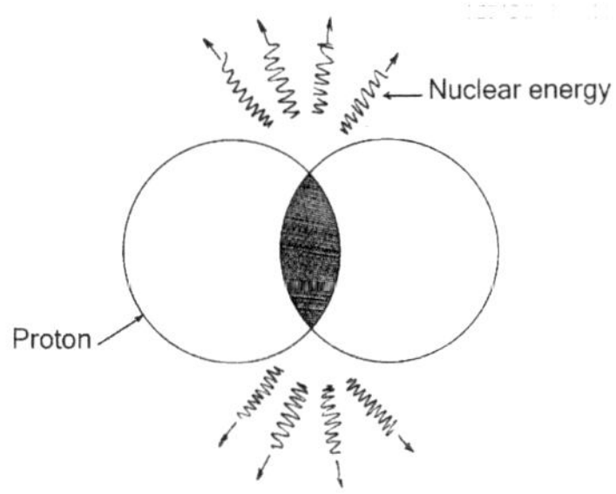

Fig 76 Nuclear energy

"...we have the formulas for that, but we do not have the fundamental laws. We know it's not electrical, not gravitational and not purely chemical, but we do not know what it is." [2]

The vortex theory can explain nuclear energy and many other things that have not been explained satisfactorily in physics and astronomy.

Proton and neutron vortices saturated with captured energy could be likened to glasses full of water. If one full glass is placed inside another, its capacity to contain water would be reduced so some of the water would spill out. In like manner, as proton or neutron vortices converge, their capacity to contain captured energy would be reduced so some of the captured energy would spill out as nuclear energy. As captured vortex energy constitutes part of the mass of a proton or neutron the loss of any of this captured energy would be associated with a marginal loss of mass. In nuclear fusion this mass loss is 0.7% of the proton mass.

During nuclear fusion, only a small fraction of captured energy is lost as nuclear energy most of it would be left inside the converged vortex particles. Swirling between them after they have converged this residual captured energy would bind them together. That could account for the strong nuclear force, which binds protons and neutrons in the nucleus of an atom. As captured energy is the cause of this force its range would obviously be limited to the nucleus.

The captured energy swirling between nuclear vortex particles, responsible for nuclear binding, would correspond to the pi-mesons or pions in nuclear particles. The vortex capture theory thus complies with Yukawa's model for

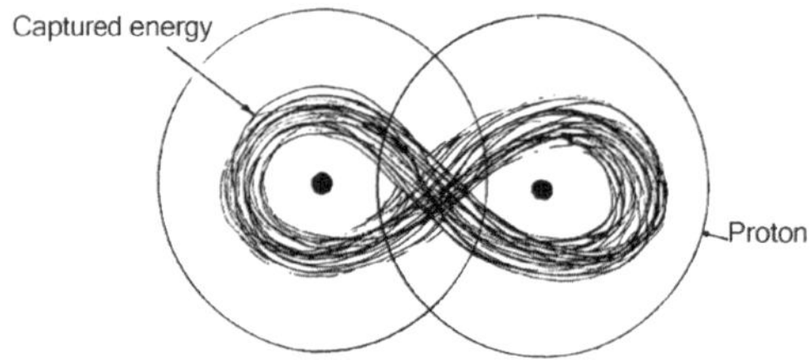

Fig 77 The strong nuclear force

the strong nuclear force as an exchange of meson energy between protons.

The vortex theory for nuclear binding can also explain why the binding between nuclear particles increases as the binding energy decreases. Uranium and plutonium have large nuclei. When these divide during nuclear fission they split into smaller nuclei. In the smaller nuclei the proton and neutron particles, called nucleons, bind together more tightly. This is because in smaller atomic nuclei the distances between the centres of the nucleons are less than they are in the large uranium or plutonium nuclei. The vortex account for nuclear fission is that as the nucleon vortices clamp together more tightly, in the formation of the smaller nuclei, they squeeze out a small amount of the captured energy, which is lost as nuclear energy. The majority of the captured energy left swirling between the nuclear vortex particles would follow a shorter circuit in the smaller nuclei. The same energy swirling in a shorter circuit would bind the nuclear vortices more tightly together. After nuclear fission the loss of a minute amount of binding energy would be marginal compared to the increase in binding between nuclear particles due to the large amount of residual captured energy racing in a tighter circuit between them. This simple account in the vortex theory for the increase in nuclear binding after nuclear fission, despite a loss of some of the binding energy, adds to the advantage of the vortex theory over the standard theory in physics.

Fig 78 The hedgehog analogy

I use an analogy of hedgehogs to illustrate nuclear energy and nuclear binding. Imagine the proton vortex as a hedgehog and the captured energy as its fleas. In much the same way that a hedgehog carries a resident population of fleas, so a proton vortex would carry captured energy. Hedgehogs don't do anything to acquire fleas, the fleas just hop onto them. So with a proton vortex it does nothing to acquire its captured energy. It is the propagating wave form of energy that drives into its spiral space path. If a hedgehog is weighed, the greater part of the measured mass would be that of the hedgehog. A lesser part would be that of the fleas. In like manner the greater part of the measured mass of a proton would be that of the stable proton vortex but a lesser part would be that of the unstable swirl of captured energy it contains. The hedgehog prickles represent the charge repulsion between protons. Because of their prickles hedgehogs converge with difficulty. So, because of their charge repulsion, protons converge only if they collide with considerable force. Imagine if two hedgehogs were forced together, as their prickles converged there would be less space between them for fleas and so some of the fleas would be evicted and hop away. Similarly, as two protons converge with force there would be less space within them for captured energy and so some would be lost and radiate away as nuclear energy. Because of the loss of some of their fleas the mass of the converged hedgehogs would be less than the sum of their weights before they were pushed together. So it is the mass of two converged proton vortices would be less than the sum of their masses before they collided. Most of the fleas would remain on the hedgehogs and, not being bothered about which back they bite, they would hop from hog to hog. In like manner the remaining captured energy in the converged proton vortices would circulate between them and bind them together.

To explain why a proton appears to be a hard corpuscular particle with a measurable diameter I use a candyfloss analogy. If you imagine a proton vortex as candyfloss, the energy it captures would be akin to strands of sugar that have, over time, filled the inner space in the floss to such an extent that it becomes a hard candy ball. When I was a boy we called hard candy balls 'gob stoppers'. The diameter of the ball would be the extent to which the floss could fill with extra sugar. And the extra sugar would add to its mass. So it is, the diameter of a proton represents the extent to which a proton vortex can completely capture energy and the captured energy contributes to its mass.

The captured energy swirling inside a proton is the pi-meson. Part of the measured mass of every proton would be the mass of the pi-meson or pion. (The meson mass is in the order of 14% of the mass of a proton. The actual mass of a proton vortex, without its captured energy, would be close to 86% of the measured proton mass.) The pion in a proton could account for its hardness, witnessed by the ricochetting of the electrons that bombarded protons in the inelastic electron-nucleon scattering experiments conducted between 1967 and 1973 at SLAC, the Stanford Linear Accelerator.

I first realised there were mesons in protons in 1966, when I saw they had been smashed out of protons in the nucleus of a silver atom in the historic cosmic-ray experiment of Cecil Powell. The experiment was more significant to me for revealing the potential of the vortex theory to explain energy-mass transformations. When I first saw Cecil Powell's groundbreaking cosmic-ray photograph of mesons, in the back of my A Level physics textbook,[3] the caption under the picture described the meson shower as a vivid illustration of the transformation of kinetic energy into mass. I realised immediately that if the meson particles of matter were formed out of the movement of the bombarding cosmic-ray particle, then mass must be a form of motion, and all other particles of matter, including those in the nucleus of the silver atom, must be similar forms of motion. The yogic vortex I had read about eighteen months prior provided me with a simple and obvious way to account for the transformation of kinetic energy into mass.

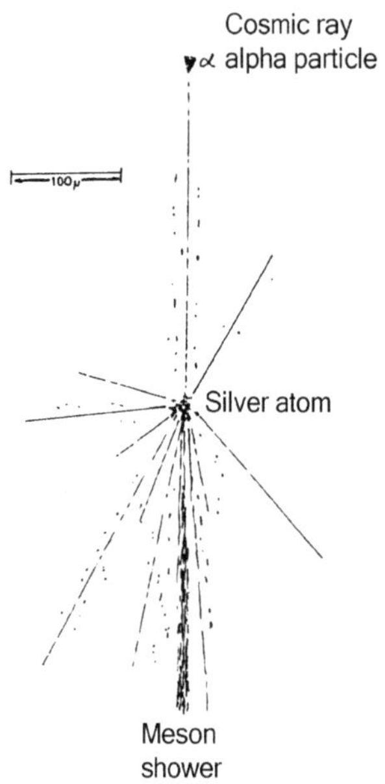

Fig 79 Energy to mass transition

A decade earlier I had learnt from my grandfather how to twist a spiral die on a soft steel rod in a vice to produce a screw bolt. That was the favourite thing I did as a boy in his workshop. Little did I know the importance of that experience because it prepared my mind to account for the results of the cosmic ray experiment. It was immediately obvious to me that if the subatomic particles of matter in the nucleus of the silver atom were vortices, they could have been effective as vortex dies through which the kinetic energy was screwed into spiral motion. When the cosmic-ray particle was stopped by the nucleus its unstoppable kinetic energy could have raced on through the proton and neutron vortices and in its passage through them would have been forced into vortex motion and so would have been transformed into mass. Years later I defined this in terms of the first quantum law of motion.

The mesons appearing on the other side of the nucleus of the silver atom were, to my mind, transient swirls of enforced vortex energy that unravelled as soon as they escaped the nuclear vortices. Their short lifespan, no more than a quadrillionth of a second, suggested to me they could have been fleeting forms of vortex energy 'caught on camera' before they reverted from vortex back to wave motion. Years later I defined this in terms of the second quantum law of motion.

The entire process of particles of energy changing from waves into vortices and then vortices back into waves, witnessed the operation of the two quantum laws of motion. In the cosmic-ray experiment 150 mesons appeared. Because there were 107 nucleons (protons and neutrons) in the silver nucleus, only 107 mesons could have been knocked out of them, to be replaced according to $E=mc^2$ from the kinetic energy of the bombarding cosmic-ray particle. 43 mesons would have been created fresh in a visible transformation of kinetic energy into mass; of wave energy into vortex energy.

In later years I used a doughnut analogy to describe the transformation of the wave-kinetic energy into mass. I described the process somewhat like batter being pressed through a doughnut making machine. Just as batter would take on the shape of doughnuts as it went through the machine so wave-form energy would be pressed into the vortex form as it passed through the nucleons in the atomic nucleus. If the doughnuts were to fall onto the floor they would immediately revert to a mess of batter. So the energy forced into the shape of the vortices would immediately revert to the wave form on exiting the nucleus.

Low-cost cosmic-ray research produced the groundbreaking results but it was haphazard so it was replaced by precision high-energy research, made possible by the construction of hugely expensive particle accelerators. The mass production of short-lived particles in high-energy particle colliders was, to my mind, nothing more nor less than the mass production of quantum doughnuts.

I was able to account for the formation of short-lived particles in high-energy laboratories like CERN and Fermilab in terms of the quantum laws of motion. According to the first quantum law of motion, as a quantum of waves passed through a quantum vortex of energy, in the process of a nuclear collision, it would take on the form of the vortex and would appear as a massive particle. According to the second quantum law of motion, on escaping from the constraining influence of the vortex of energy, the energy in the transient mass of the vortex form would immediately revert to wave-form motion and radiate away.

As physicists increased the energy in their experiments they produced ever more massive particles. This is because increasing the energy of the accelerated particles increased the transient mass of the temporary particles produced. The second quantum law of motion accounted for the extremely short life span of these synthesised particles.

However, in the high-energy laboratories, physicists were discovering particles that were lasting longer than expected. The difference in lifespan was a thousandfold. Instead of lasting for only a trillionth of a second they were surviving for a billionth of a second. Murray Gell-Mann, author of the quark theory used in physics to explain the outcomes of high-energy experiments, called these new particles 'strange' for their exceptional longevity. He went onto speculate the existence of strange quarks to account for them and designated strangeness as a new property of matter.

The quantum laws of motion provide a straightforward way to explain strangeness. When strange particles decay they leave a proton or an electron behind. It could be that when wave form energy is blasted through an atomic nucleus, in a high-energy experiment, an unstable vortex sometimes forms around a stable electron or proton vortex in the nucleus. Should this occur the stable electron or proton vortex could have a scaffolding effect on the unstable vortex, conferring upon it a degree of stability that would increase its lifespan. The scaffolding effect would delay the action of the second quantum law of motion. The transient vortex energy swirling around the stable seed vortex would be slowed by the scaffolding effect in its process of reverting back to its natural wave form.

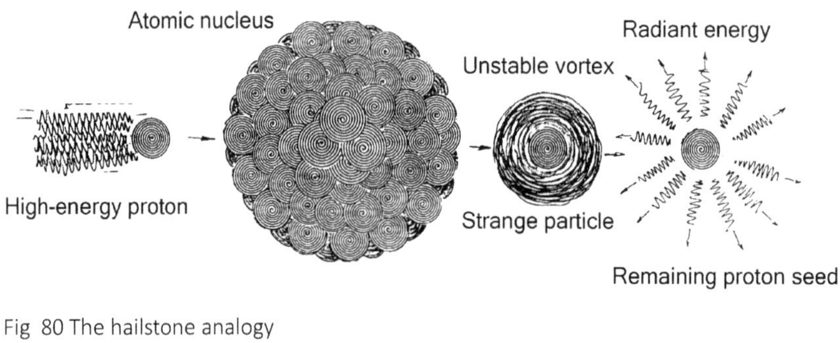

Fig 80 The hailstone analogy

As physicists increased the energy in their accelerators, they found some of the massive particles were generated in layers around a natural particle. I imagined this occurring much like ice forming in layers around a particle of dust in a hailstone. I used a hailstone analogy to explain how some of the newly-formed massive particles decay in steps, or layers, to reveal a new, lighter particle at each lower level of energy where particles tended to appear. This cascade decay of unstable particles fits with the model of a hailstone melting, layer by layer, in reverse of the way it formed. The levels of mass-energy forming round a proton in the creation of a strange particle were called by Gell-Mann 'doses of strangeness'. The decay of the strange particles in steps, to reveal lighter particles on each step, was described as shedding doses of strangeness.

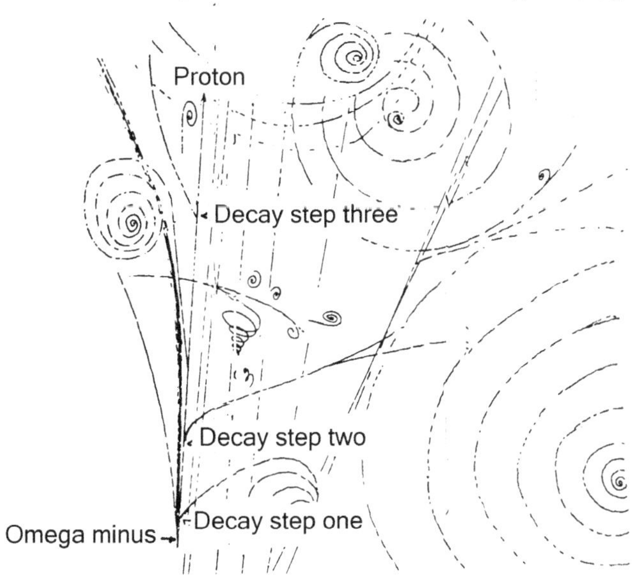

Fig 81 Cascade decay of the Omega minus

According to the vortex theory, as the cascade decay occurred and each dose of strangeness was shed it left a less massive vortex particle behind. The scaffolding effect of the proton vortex over the residual transitional mass would then be stronger because there would be less swirling energy to stabilise, and it would be closer to the stabilising proton vortex at the centre of the swirl. This increase in the strength of the scaffolding effect meant that as each dose of strangeness was shed the particle left behind would last longer than the one that went before it. This account for shedding doses of strangeness in the vortex theory is evident in the cascade decay of a particle called the Omega minus. In February 1964, at the Brookhaven National Laboratory, Nicholas Samios photographed the cascade decay the Omega minus.[4] If you look closely at the image of the three-step cascade decay of the Oomega minus, you will notice that the distance between each successive step is longer in the progressive cascade decay. The length of the track a particle leaves before decaying gives a measure of its lifespan. This shows that each residual particle in the cascade decay process lasts longer than the particle before it, precisely as expected in the vortex theory.

The fact that strange particles sometimes form around electrons confirms that there are electrons in the nucleus of the atom. This adds to the evidence against Heisenberg's uncertainty principle because it could be argued that the electrons originate from neutrons in the nuclei. Where else would they come from? Strangeness adds strength to the contention in the vortex theory that neutrons are electrons bound to protons.

1. Einstein A., Infeld L., *The Evolution of Physics,* Simon & Schuster, 1938
2. Feynman R., *The Feynman Lectures on Physics, Vol 1.*, Addison Wesley, 1963
3. McKenzie A. E., *A Second MKS Course in Electricity*, Cambridge University Press, plate 19, 1968
4. Calder N., *Key to the Universe: A Report on the New Physics* BBC Publications 1977.

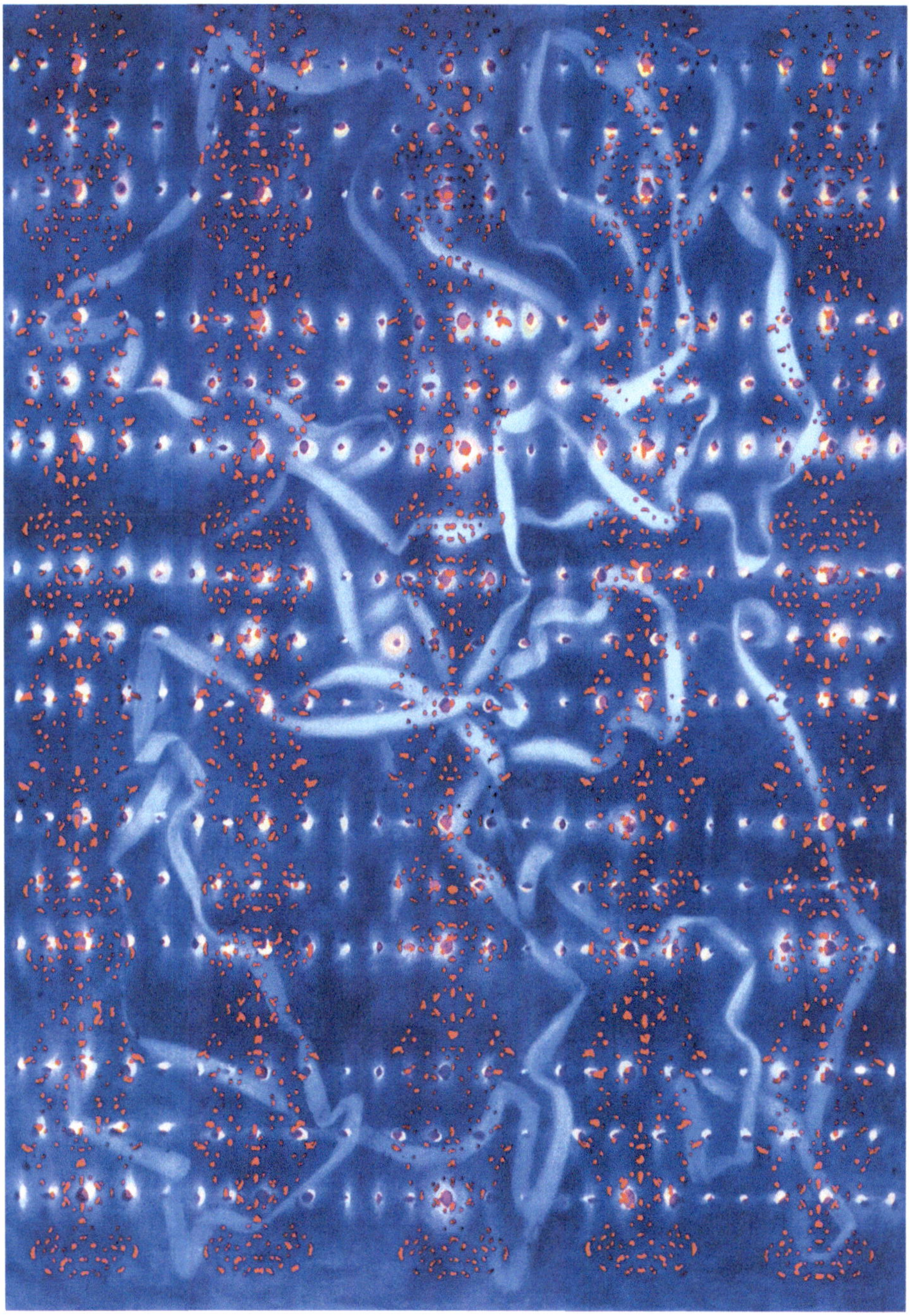

9. Sainted Hunger - Shane Guffogg

"In beta-decay electrons are observed to come away from neutrons with a range of energies." - Ash

ESSAY 8

Neutrons and Neutrinos

"These fundamental things have got to be simple"

Ernest Rutherford

Précis In the vortex theory, neutrons are treated as electrons bound to protons by the electric force of attraction. If the electron and proton are held together in the neutron in this way, why do neutrons decay so easily outside a normal atomic nucleus or in radioactive nuclei? The answer to this question is that the kinetic wave train of energy captured by an electron vortex acts against its electric attraction to the proton vortex.

Like a ping pong ball bouncing on a fountain of water, it is only a matter of time before the electron breaks free. According to the vortex theory, in the neutron the kinetic wave energy attached to the electron vortex swirls in the proton vortex adding to the proton's captured energy, which causes the proton to be in a state of energy supersaturation. This pushes against the electron vortex like water in the fountain. As the electron bounces free it leaves some of this energy behind which is then lost as a neutrino as the proton normalises from supersaturated to saturated. In the vortex theory, the neutrino is not considered to be permanent as most physicists like to believe; it is a momentary swirl of escaping captured energy that rapidly reverts to radiant energy. If a proton is imagined as a quantum cat and an electron is imagined as a quantum mouse, the neutron could be thought of as cat with a mouse in its jaws. Think of the tail of the quantum mouse tangled in the teeth of the quantum cat. As the lively quantum mouse escapes, a bit of its tail breaks off and is left in the quantum cats mouth. It doesn't stay there but falls out and is lost. That analogy gives account for the neutrino.

In the vortex theory, the neutron is not thought to be stable in the nucleus of an atom. The nucleus can be imagined as a group of quantum cats. Some of them have quantum mice in their mouths. These are the neutrons. The quantum cats without mice represent protons. As soon as an electron mouse escapes from the jaws of a proton cat, a proton cat without a mouse pounces on it. The result of this situation is that the mice are unlikely to escape from the group of cats. This analogy represents an idea in the vortex theory that, in the nucleus of atoms, electrons are continually exchanged between protons. This can explain radioactivity.

Radioactive atoms represent large or small groups of cats from which mice are more likely to make an escape. Very large groups of cats represent large radioactive atomic nuclei of elements like uranium. Sometimes a quantum mouse will bolt from the group. That represents an escaping electron. Sometimes a small group of quantum cats with a few quantum mice between them will come tumbling out and roll away. Sometimes a cat with a mouse in its jaws will break free of the group. That can cause trouble in uranium atoms.

One particular uranium house is very unstable. It has a quantum cat flap through which every now and then a cat with a mouse in its jaws gets kicked out. It then nips through the cat flap into another uranium house. That is one cat and mouse too many in an already unstable house. The uranium nucleus swells then, in a spat of hissing and spitting feline fury, it splits releasing fire and two or three cats with mice. These nip through the cat flaps of two or three more uranium houses. That can set up a chain reaction leading to a nuclear explosion as depicted in the movie, *Oppenheimer.*

~

Fig 82 The ping pong analogy

Essay: In the free state neutrons have a half-life of eleven minutes. This means after eleven minutes half of the neutrons in a group of them will have disintegrated into an electron and proton, and in the next eleven minutes their number would drop to a quarter. To understand this imagine a fountain of water with a ping pong ball bouncing on it. In this analogy the ping pong ball would represent the electron and the fountain would represent energy in the neutron as a counter force acting against the electrostatic binding of the electron to the proton. Now imagine you are watching a number of fountains with ping pong balls bouncing on them. Balls are breaking free of the fountains at random. You record the time it takes for half the balls to escape their fountains. It is eleven minutes. Then you record how long it takes for half the remaining balls to bounce free and you find it is still eleven minutes. The time interval between the numbers of balls left on the fountains between a half and a quarter is the same as between the whole and the half. This is a feature of radioactive decay called half-life. In the ping pong analogy the half-life of the balls bouncing on fountains would be eleven minutes.

If a neutron is a bound state of an electron and a proton caused by opposite charge attraction why would neutrons be unstable. Why after a few minutes would they fall apart? There must be a counter force in the neutron acting against the electric force of attraction between the electron and the proton. Could it be the energy caught up in the mass of the neutron exceeding that of an electron and a proton? This has been measured as the mass equivalent of one and a half electrons according to $E=mc^2$. Compared to an electron this would be an enormous amount of energy. It would have been the kinetic energy of the electron before it became attached to the proton in the neutron. This energy could be acting against the electrostatic attraction between the electron and proton that binds them together in a neutron. It could be responsible for its instability. According to the vortex theory when the momentum of the electron was arrested by the proton in the formation of a neutron, its kinetic energy would have driven on into the proton and adding to the energy in complete capture it would increase the proton mass.

In the vortex theory, the stable mass of a proton is the mass of the proton vortex plus the maximum amount of energy it can contain in captured energy. This could be described as its 'capture saturation mass' represented by the mass of a pi-meson or pion it contains. The partial captured energy brought in by the electron, when a neutron is formed, would cause the proton to become supersaturated with captured energy. This state of supersaturation could set up instability in the neutron. It might be only a matter of time before the excess captured energy would overcome the electrostatic attraction and explode the electron out of the neutron causing it to break free of the proton. In the ping pong analogy this state of supersaturation could be imagined as the pressure of the water in the fountain pushing against the ball.

The pressure of excess mass in the proton could cause the decay of the neutron known as beta-decay. In beta-decay electrons are observed to come away from neutrons with a wide range of energies. These energies are somewhere between the minimum required to overcome the electric attraction between proton and electron, and the maximum reflected in the mass of the neutron exceeding the sum masses of an electron and a proton. The varying energies of electrons exiting neutrons suggests some of the captured mass of supersaturated energy is left behind by the electron as it rips away from the proton. To restore the saturation level of captured mass in the proton this remnant mass of supersaturation would have to break away from the neutron at the same time as the electron is emitted in beta-decay. Because electrons leave neutrons with a range of energies the exiting remnant mass would vary from one neutron decay to another. According to the vortex theory this remaining fragment of short-lived transitional mass, escaping complete capture, would be a momentary swirl of energy with random mass reverting from vortex to wave form before it radiates away. It would seem when an electron explodes from a neutron it doesn't redeem all the kinetic energy it came in with when the neutron was formed. Some is left behind and leaves the proton independently as the neutron decays.

The remnant mass escaping the neutron alongside the electron in beta-decay proved difficult to detect. In 1930, Wolfgang Pauli suggested that it left the neutron along with the electron as a discrete particle. When he predicted it he is reported to have said: "I have done a terrible thing. I have postulated a particle that cannot be detected". In 1933, Enrico Fermi called Pauli's little particle a 'neutrino' meaning in Italian 'the little neutral one'. The neutrino proved to be very nebulous. It was extremely light, electrically neutral and did not participate in any force. It was also very short ranged and so was very hard to find. It could only be inferred from the interactions of other particles until Frederick Reines headed a team that managed to identify neutrinos in beta-decay. He published the finding in 1956 [1] but didn't receive a Nobel Prize for the achievement until 1995. In 2015, Takaaki Kajita and Arthur McDonald received the Nobel Prize for determining that neutrinos have mass even if they were unable to measure it with precision. This was obviously because the masses of neutrinos are imprecise.

Fig 83 Wolfgang Pauli

The mass of neutrinos would be difficult to determine if they represent a random amount of energy escaping the neutron alongside a beta particle electron. Because the beta electrons leave neutrons with variable energies the energies of neutrinos would be infinitely variable. The neutrino would be, in effect, a random fragment of pion mass spat out of the neutron as it decays so neutrinos would not have a consistent, predictable mass like most other short-lived particles emerging in high-energy research. Compounded by their electrically neutrality and lack of interaction with the strong nuclear force that would make them extremely difficult to detect.

In standard physics, neutrinos are conjectured as long-lasting natural particles but the premise in the vortex theory that they are just rump fragments of pion decay is supported by a recent astronomical observation of neutrinos in the central rim of the milky way [2] where neutrinos were discovered to have been generated by decaying pions.

This discovery adds weight to the vortex theory that pions are unstable vortices of energy escaping capture and that neutrinos are just small fragments of escaping captured energy reverting from vortex to wave motion. Neutrinos may not be significant particles. They may be just random scraps of vortex energy reverting to their natural wave form after escaping capture. Along with decaying pions, neutrinos may be just observations of the operation of the second quantum law of motion.

A question remains: Why are neutrons stable in atomic nuclei? I don't believe they are. I believe they only appear to be stable in an atomic nucleus because in the event of an electron escaping a neutron it is liable to be caught by a proton, so that the proton and neutron effectively swap identities.

Apart from hydrogen, which is one proton orbited by one electron, most atoms contain neutrons as well as protons in the nucleus. In most of the lighter atoms the ratio of protons to neutrons is 1:1. As the nuclei become larger with more protons, the ratio increases to 1.5 for the heaviest stable nuclei. This represents an increase in the number of neutrons contributing nuclear binding, without adding to the charge repulsion between protons. My father calculated the ratio of protons to neutrons in the nuclei of atoms and worked out if the ratio exceeds three neutrons for every two protons the nucleus becomes unstable. This I believe is because while neutrons contribute to the stability of the nucleus by offsetting repulsion between protons they introduce instability through the electrons they contain.

I believe that in atomic nuclei, electrons are exchanged between protons and neutrons. If an electron breaks free of one proton it could be caught by another. The neutron to proton ratio of 1:1 in smaller nuclei and 2:3 in larger nuclei would seem to present the greatest likelihood of this occurring. A stable nucleus of carbon contains six protons and either six or seven neutrons. However there are some nuclei of carbon that contain eight neutrons. They are

Fig 84 Michael Ash

unstable because the proton to neutron ratio is 2:4. These atoms of carbon are radioactive. It is only a matter of time before a beta-decay electron will escape the nucleus leaving it with seven protons and seven neutrons. The half-life of radioactive carbon is about 5,700 years, so the age of an old wooden artifact can be measured by comparing the amount of radioactive carbon it contains with the radioactive carbon in a new piece of wood. I explain radioactive decay with a cat and mouse analogy.

Imagine electrons in the nucleus of an atom as mice and the protons as cats and the tails of the mice as the kinetic energy of the electrons. An atomic nucleus, with its mix of protons and neutrons, would be a represented by a group of cats in which some have a mouse in their jaws or claws and some are without a mouse. If a mouse breaks away from a cat it would be caught immediately by one of the cats without a mouse. So long as there is a balance of at least two cats for every mouse it is unlikely a mouse would ever break free from the group.

The nucleus of a radioactive carbon atom would be represented by a group of fourteen cats with eight mice between them. Six of the cats are without a mouse. They represent six protons in the nucleus. Eight cats have a mouse each. They represent the eight neutrons. The most stable ratio to ensure a mouse doesn't escape is a cat without a mouse watching a cat with a mouse. The watching cat would be ready and waiting to pounce as soon as the mouse escapes. The perfect balance is six cats with mice and six without. Seven cats with mice is still within the manageable ratio but eight cats with mice, watched by only six without mice, is a recipe for disaster. There simply aren't enough vigilant cats to catch every single escaping mouse. It is only a matter of time before a mouse breaks free of the group leaving seven cats with mice and seven without. Now every mouse in jaws is watched by a claw ready cat set to pounce on it if it breaks away.

A cat on its own with a lively mouse would stand more of a chance of losing it than if it were in a congregation of cats continually pouncing on mice as fast they try to escape. This exemplifies why neutrons in the nucleus of an atom appear to be more stable than neutrons in a free state.

However, the larger the group of cats, the less likely it would be to have an optimum balance between cats with mice and cats without mice. In that situation there might be a better chance for a determined mouse, or even some cats with mice, to break free. This is how the cat and mouse analogy can be used to explain radioactivity.

Atoms, like uranium, with large, loosely-bound nuclei, tend to be radioactive. They are represented by large groups of cats in the cat and mouse analogy. Radioactivity comes in the forms of high-energy gamma ray photons, beta-decay electrons, represented by mice in the analogy, and alpha-decay helium nuclei, represented by a four cats with two mice continually tossed between them. The loss of a neutron from a nucleus is represented by a single cat breaking out of a group with a mouse in its jaws.

Fig 85 The cat and mouse analogy

Escaping mice pay a price. As a mouse breaks free of a cat it is liable to lose a bit of its tail to tooth or claw and come away with a random length of remaining tail. That is how I use the cat and mouse analogy to explain the escape of electrons from radioactive atoms, which come away in beta-decay with a range of energies. The remnant bit of tail of random length, which falls away from the mouse as it breaks free, represents the neutrino with its random mass.

I also use the cat and mouse analogy to explain the detonation of nuclear weapons. Because atoms in the same element can vary in the number of neutrons they contain, the term 'isotope' is used. Uranium (U), for example, has three naturally occurring isotopes, U^{234}, U^{235} and U^{238}. Each isotope of uranium has the same number of 92 protons but an atom of uranium^{234} has 142 neutrons; an atom of uranium^{235} has 143 neutrons and an atom of uranium^{238} has 146 neutrons. They are all radioactive because the proton to neutron ratio in them all is above 1.5. Uranium^{238} is the most common isotope. Uranium^{235} is less plentiful and uranium^{234} is the rarest natural isotope as it is the breakdown product of uranium^{235}. This breakdown occurs because U^{235} can lose a neutron. It is this discovery that led to the atomic bomb.

Uranium^{235} (U^{235}) is the fissile isotope of uranium that was used in the Oppenheimer atomic bomb. If a neutron enters its nucleus the nucleus will split in two and release two to three more neutrons. In my cat and mouse analogy I depict U^{235} as having a metaphorical cat flap that lets in stray cats looking for a home. However, when a stray with a mouse in its jaws, enters the cat flap of a $uranium^{235}$ home it so disturbs the equilibrium of the cats already there that they fall out bigtime. The congregation of cats first becomes turbulent, then it swells and finally it splits in two, releasing pent-up feline fire and fury from the group. This represents the release of nuclear energy as U^{235} splits. As it does so, two or three cats, each with a mouse in its jaws, break free. These awful jawful strays then look for new homes and if they find more of the $uranium^{235}$ atoms, each enters an allegorical cat flap uninvited and splits the resident group, releasing more furious felines with a lot more hissing fire and spitting fury.

Fig 86 The Einstein letter

Physicists working on the Manhattan Project – initiated by a letter signed by Albert Einstein on 2 August 1939 and delivered to President Roosevelt on 11 October – isolated, purified and concentrated $uranium^{235}$ and then crushed it into a 15kg ball. As they did so a chain reaction began. Neutrons, released from spontaneous decays of U^{235} to U^{234}, entered more U^{235} nuclei and split them to release two or three neutrons. Each one then went on to split two or three more U^{235} nuclei causing the release of four or six more neutrons, which then split another four to six U^{235} nuclei to release eight to twelve neutrons then sixteen to twenty-four, thirty-two to forty-eight, and so forth. This set up the chain reaction which rapidly broke up ever more nuclear households. With the split of each U^{235} nucleus, into the smaller tighter nuclei of iodine, caesium, strontium, xenon or barium, nuclear energy was released and in an exponential flash, as the number of splitting nuclei grew to countless trillions, a nuclear explosion occurred.

Fig 87 Nuclear explosion

In the chain reaction over Hiroshima, thirteen-thousandths of a second was all the time it took for neutrons to penetrate and split the nuclei of practically all the $uranium^{235}$ atoms in the 15kg ball. This caused the explosion of the first nuclear weapon, which destroyed in seconds the entire city and practically all of its inhabitants, leaving survivors with long term debilitating radiation sickness, cancers and other diseases as well as DNA damage that passed ongoing disability and chronic ill health onto future generations.

1. Cowan C., Reines F., et al, *Detection of the Free Neutrino: a Confirmation*, Science, July 20, 1956
2. Pappas S., *In a First, Scientists See Neutrinos Emitted by the Milky Way*, Scientific American, June 29, 2023.

10. A Rose is a Rose is a Rose - Shane Guffogg

In this beautiful image I visualise quantum entanglement

ESSAY 9

Was Newton Right?

"The highest wisdom has but one science, the science of the whole"

Leo Tolstoy

Précis: There is a common assumption that Einstein toppled Newton, but in the vortex theory Newton and Einstein are reconciled. Einstein developed his theory of gravity from a thought experiment of a man in a box. If the box were to free fall due to gravity, as it accelerates the man would be weightless so the acceleration would cancel out the pull of gravity on him. If the box were to accelerate in space, where there is no gravity, the man would fall toward the back of the box as though he was pulled by gravity; so, Einstein argued, gravity could be cancelled or caused by acceleration. He based his theory for gravity on acceleration. Newton based his theory for gravity on an extraneous force field.

If you consider two interacting vortices of energy, viewed from the outside they would appear as overlapping fields of force. That represents Newton's point of view. If you were to view the vortices from inside, relative to each other, they would appear to be accelerations. That is because energy accelerates into the vortex. This represents Einstein's point of view. As it is impossible to devise an experiment to prove one point of view is truer than another, Newton and Einstein would both be right if gravity is caused by a vortex interaction.

The fourteenth century philosopher, William of Occam, is famous for stating: "Entities are not to be multiplied beyond necessity". He argued that, in a choice of theories, we should accept the one that is simplest and has the least arbitrary assumptions. He argued this law of economy of ideas with such sharpness that it became known as Occam's razor. When it comes to a choice between the force-carrying graviton theory, Einstein's theory, Newton's theory or the vortex theory, the vortex theory scores because it is simple, straightforward and incorporates elements of both Newton and Einstein's theories; and, as well as gravity, it explains practically everything else in physics from the single idea that subatomic particles are vortices of energy.

Blackholes are the greatest concentration of vortex energy. They are the densest centres of matter in the Universe. There, gravity is strongest. If gravity is caused by a pull between matter and antimatter through the centre of everything, then, through the centre of blackholes, annihilation between matter and antimatter should occur. A feature of blackholes is that their gravity if is so strong it traps light. However, if annihilation energy is constantly flowing through the centre into the blackhole, eventually the amount of energy would overcome the ability of the gravity to hold it and, in the manner of geyser, gamma ray energy – the hallmark of matter-antimatter annihilation – would periodically burst out of the blackhole. That is how the vortex theory accounts for periodic gamma ray bursts from the blackhole centres of distant galaxies. This demonstrates the power of the vortex theory not only to explain practically everything in physics but also to explain things that are as yet unexplained. According to the vortex theory, the Universe may not be endlessly expanding as most physicists believe. The cause of distant galaxies accelerating away from us could be the polar opposite of gravity where galaxies, the largest things, are pulled toward the largest sphere of space where they may meet and annihilate antimatter galaxies. At both ends of the size spectrum of space the vortex reverts to light, through matter-antimatter annihilation. As stars die and are absorbed by blackholes they are cremated by annihilation but, with mass, space is also lost so compression would occur. That sets up for another Big Bang.

~

Fig 88 Newton

Essay: On 7 November 1919 The Times of London published a headline: REVOLUTION IN SCIENCE New Theory of the Universe NEWTONIAN IDEAS OVERTHROWN. Three days later The New York Times followed with: LIGHTS ASKEW IN THE HEAVENS Men of Science More or Less Agog Over Results of Eclipse Observations, and EINSTEIN THEORY TRIUMPHS. The newspapers created a media storm. Einstein became a sensation, but did his theory for gravity triumph over Newton's?

Were Newton's ideas really overthrown?

Einstein based his general theory of relativity on a principle he developed from his famous thought experiment of a man in a box. If the box were to free fall from the sky the man would float in the box and no longer experience the force of gravity. If the box were to accelerate upwards in a gravity free environment the man would fall toward the floor of the box as if it was the force of gravity. Einstein argued that acceleration could mimic or cancel out the force of gravity and the man in the box would have no idea whether he was experiencing the force of gravity or an acceleration. In this, The Principle of Equivalence of Forces , Einstein declared it impossible to devise an experiment that could prove a force is either due to a force field or an acceleration because, he argued, these were just different points of view.

In 1972, when I was developing the vortex theory for gravity, I realised that in my thesis Einstein's principle of equivalence of forces did not allow for Einstein to topple Newton with his general theory of relativity. It might have appeared he dismissed Newton's force field theory of gravity but his principle did not allow for his acceleration theory to be closer to the truth than Newton's as they appeared to be just two points of view of a vortex interaction.

I came to this realisation in a recording studio off Holland Park avenue in London. I was transfixed by reels fast winding on the recording decks. I saw the spools accelerate as the tape fast wound off them in a spiral toward their centres. In a flash I saw vividly how spiral motion toward the vortex centre would result in an acceleration down its radius. The studio was walking distance from the college library where I had read about Einstein's principle. My life was full of amazing coincidences. I was led from step to step in my work. The way I was drawn to the book that contained Einstein's principle was amazing. I was in the library of Queen Elizabeth College, adjacent to Holland Park, revising, when I felt an urge to get up and go to a row of books. A small green book stood out. I felt a wave of excitement as I picked it from the shelf. It was on Einstein's theories of relativity. I opened it at random. The page I had opened was on Einstein's Principle of Equivalence of Forces. In the recording studio my family were excited about the songs we had just recorded. I was in another world. As I watched the spinning spools all I could think about was Einstein's principle.

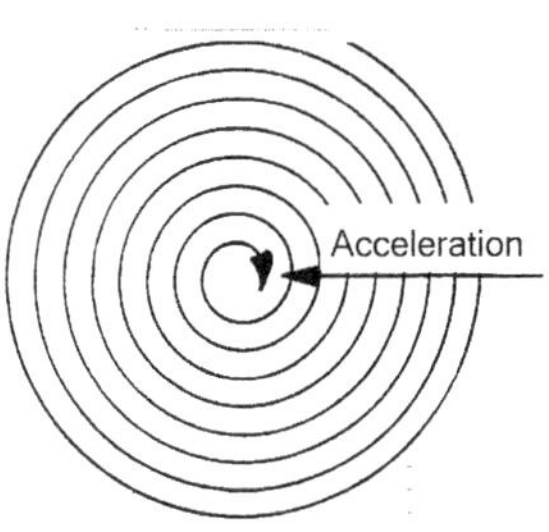

Fig 89 Acceleration in the vortex

When spheres are placed in contact they touch at the point of a radius. As

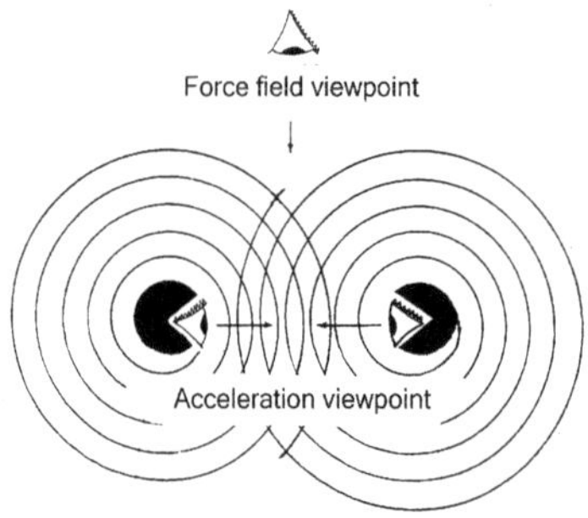

Fig 90 The Principle of Equivalence of Forces

a vortex of energy is effective as an acceleration down the radius toward its centre, from the perspective of observation of one overlapping vortex of energy relative to another they would appear to be an acceleration. This represented Einstein's point of view. However, from the perspective of an observer looking down on a pair of overlapping vortices of energy they would appear as interacting force fields. This represented Newton's point of view. As no one can determine experimentally whether one point of view was more correct than the other, the theory for gravity based on vortex interactions would accommodate both Einstein's and Newton's points of view about gravity. The vortex theory reconciles Newton and Einstein, and in fact it is more Newtonian than Einsteinian.

The Royal Society observation of the apparent deflection of starlight around the Sun was taken to confirm Einstein's theory. However, the 1919 observation could equally confirm the view that concentric spheres of vortex energy extending from the Sun caused the starlight to be deflected. This view would support the model of a force field if the concentric spheres of vortex energy extending from the Sun were to transmit the Sun's gravity into space. If the vortex theory could show gravity was caused by a vortex interaction, it could also show the effect could be transmitted into infinity by the Sun's extending field of space.

I was working on my matter-antimatter model for the Universe in 1972 when I realised there could be a connection between the force of gravity and antimatter. This was because the force of gravity came from the centre of matter and, in my model, antimatter occurred beyond the centre of every subatomic particle of matter in existence. The nuclear physics I learnt from my father when I was a boy was based on the work of Rutherford. This is why my thinking of a vortex particle of antimatter with an opposite charge, beyond the centre of every charged vortex particle of matter, was based on Lord Rutherford's Law of Conservation of Electric Charge, which declared that for every charge in the Universe there is an equal and opposite charge.

Particles of matter and antimatter are opposite in charge and between opposite charges there is an accelerating force of attraction. In the recording studio my mind was racing. I realised antimatter could account for the accelerating force of attraction we experience as gravity. This is when I realised if there is antimatter beyond the centre of matter then a vortex interaction acting through the collective centres of subatomic particles could cause them to centralise on each other, which is the hallmark of gravity. This centralising pull of gravity, acting through the centre of matter, would be transmitted through the entire vortex extending into infinity. If you pull a person by a finger, the person will follow the pull because the finger is an integral part of that person. Likewise, the spheres of space extending from the Sun would transmit the Sun's gravity into infinity because they are an extension of the Sun; and gravity and space have a lot in common.

	Gravity		**Space**
1	Extends from matter	1	Extension of matter
2	Acts from the centre of each particle of matter	2	Extends from the centre of each particle of matter
3	Unlimited range in 3D	3	Unlimited extension in 3D
4	Obeys Inverse Square Law	4	Obeys Inverse Square Law
5	Curves the path of light	5	Curves the path of light
6	Universal gravity is the sum of the gravity acting from every particle of matter	6	Universal space is the sum of space extending from every particle of matter

Einstein's general theory of relativity is a story for gravity. The vortex theory is a different story. Both stories account for the deflecting effect that massive bodies, like the Sun, have on starlight. Einstein's story is built on an arbitrary assumption that the mass of the Sun can curve spacetime. The quantum vortex theory, that space is curved because it is an extension of the shape of the Sun, is a direct outcome of the single assumption that subatomic particles are spherical vortices of energy. Because of this Occam's razor can be invoked for priority of the vortex theory over Einstein's theory.

William of Ockham or Occam was a fourteenth century philosopher famous for stating: "non Sunt multiplicanda entia praeter neccessitatem – Entities are not to be multiplied beyond necessity". He argued that in a choice of theories we should accept as closest to the truth the one that is simplest and has the least arbitrary assumptions. He argued his law of economy of ideas with such sharpness that it became known as Occam's razor.

In the vortex theory, gravity is derived from a unified description of space, mass and the force of electric charge as aspects of the vortex of energy. Because the quantum vortex is formed of movement at the speed of light, space, time and mass are relative to this invariable speed. Because the quantum vortex is curved, space is curved. Because energy in the quantum vortex extends beyond our direct perception, bodies of matter can act on each other at a distance. Because the quantum vortex is intrinsically dynamic these interactions can occur without the mediation of unknown entities.

In the vortex theory, the extending force fields envisioned by Isaac Newton are interacting quantum vortices. No arbitrary assumptions are layered on the single axiom that subatomic particles are vortices of energy. In the theory of the quantum vortex the forces and properties of matter and antimatter, including gravity and space, are accounted for in a single, simple process of unfolding logic.

In 1972, I realised I could also use the vortex theory for gravity to explain blackholes. The premise for blackholes came from reasoning that if quantum vortices of energy are particles of space that can overlap and occupy each other then it is conceivable they could all converge as a single vortex. In the vortex theory, it is feasible that the entire Universe could collapse, due to gravity, onto a singularity point less than the size of an atom, but the electric force of repulsion between protons would work against gravity to prevent this collapse from happening. Protons packed in the nucleus of an atom do not converge into singularity because the force of electric charge pushing them apart is stronger than the force of gravity pulling them together. However, in the demise of stars, when the loss of hydrogen fuel reduces the kinetic energy of electrons, the electrons can fall into protons to form neutrons. Then with the electric force cancelled out, the force of gravity could take over. That is when the collapse of matter into singularity would be possible. In this process small stars could become dwarf stars and more massive stars could collapse into neutron stars. The most massive dying stars would become blackholes. Representing the greatest density of matter, blackholes are the greatest concentration of space in the Universe.

According to the vortex theory when stars in matter collapse toward singularity the same would be happening to their antimatter counterparts and, as they converge, annihilation would take place. This led me to conclude that the energy of annihilation could be released by the action of gravity and the release would occur through the singularity points in the core centres of black-holes. Quasars – quasi-stellar phenomena – had been recently discovered and I thought they might be evidence of this. [1]

When they were discovered quasars were then thought to be the most energetic things in the Universe. Releasing nearly two hundred times the energy of stars, their energy profile fitted matter-antimatter annihilation. (The energy release of nuclear fusion in stars comes from the burn of 0.7% of the mass of a proton so, in matter-antimatter annihilation, 100% of the mass of a proton is consumed, which releases around two hundred times the energy of nuclear fusion. Quasar emissions, therefore, can only come from the annihilation of matter and antimatter.)

Quasars have been discovered at the centres of galaxies where supermassive blackholes are found. This is where I expected them to be if their energy came from matter-antimatter annihilation through the smallest realms of space. But quasars have also been discovered at the furthermost reaches of the Universe, which suggested to me that annihilation was also occurring across the largest sphere of space. This I expected because that was where the vortex theory also suggested matter and antimatter could meet and annihilate. Then gamma ray bursts from the cores of supermassive blackholes were discovered. They added strength to the vortex cosmology as gamma rays are the hallmark of matter-antimatter annihilation.

Gamma ray bursts originate from outside our galaxy and are a thousand times more luminous than quasars and one hundred quadrillion times more luminous than the Sun. More recently they have been confirmed to be associated with the transition of massive stars into blackholes and subsequently periodic gamma ray bursts were discovered to occur from blackholes. Now they are most commonly associated with the core collapse of very massive stars into blackholes. [2].

Gamma ray bursts provide evidence that antimatter exists beyond the core centres of stars as predicted in the vortex theory and that gamma rays are released from blackholes where mass is most centralised. It is abundantly clear that matter-antimatter annihilation occurs where gravity is strongest. There can be no doubt that antimatter exists beyond the centre of matter. But gamma ray bursts occur periodically. Why would that be? In the vortex cosmology the answer is that blackholes could be behaving somewhat like Old Faithful in Yellowstone Park. If the theory of the quantum vortex is correct they should release gamma rays in the manner of a geyser. Let me explain. In the vortex theory, gamma rays would be produced continuously by the annihilation of matter and antimatter through the centre of a blackhole. In line with the first quantum law of motion, the blackhole would act as a gargantuan proton vortex holding the gamma ray photons in complete capture. Initially they would not be free to escape but with the continual build-up of annihilation energy there would come a point when the amount of energy in capture would exceed the ability of the blackhole gravity of the blackhole to hold it. Then, in line with the second quantum law of motion, the annihilation energy would escape from capture as a burst of gamma radiation after which the cycle of capture would begin all over again. I called this process the gravity geyser.

Recent discoveries of high-speed streams of electrons from the centres of galaxies support the idea that annihilation of matter and antimatter is going on inside the blackholes located within them. William Tiller of Stanford University and William Bonner of London University strengthened the quantum vortex cosmology when they related this to gravity and hidden properties of a mirror world. [3]

The evidence of antimatter beyond the core of blackholes is accumulating. It is provided by bursts of gamma rays, high-energy electrons and X-rays now recorded regularly as astronomical events. Increasingly this is supported in the scientific journals. In a Scientific American article, 'The Brightest Explosions in the Universe', [4] account was given of annihilation between matter and antimatter through the centres of blackholes.

A theoretical model in physics is urgently needed to account for antimatter beyond the centre of matter. The vortex theory not only does that but as a theory it works only if there is a particle of antimatter beyond the centre of every particle of matter. The existence of antimatter beyond the centre of matter is essential to the working of the theory because it completes the universal cycle of vortex energy necessary for the theory to comply with the first law of thermodynamics, that the total energy of a system remains constant.

Current theories in cosmology are bankrupt because cosmologists have dispensed with antimatter. There is none in their ledger to account for these events. Leading cosmologists believe that at the Big Bang there was matter and antimatter but slightly more matter than antimatter. In universities physicists and astrophysicists teach as fact that immediately after the Big Bang all the antimatter annihilated with most of the matter and the remnant is what we see as the Universe today. [5] This is an example of the fictions being taught as fact in universities. The pessimistic assumption of an indefinite expansion of the Universe may also be unfounded. The Universe may not be expanding. What appears to be an accelerating expansion of the Universe may be galaxies of matter racing out to meet galaxies of antimatter toward mutual annihilation.

In 1995, the vortex cosmology was published which included the prediction of equal antimatter to matter in the Universe. [6] Predictions of an antimatter half of the Universe go back to Paul Dirac and the theory of a Trouser Leg Universe, that came from Tel Aviv University in 1970. These proposals were dismissed for lack of evidence of gamma rays, which is the signature of matter-antimatter annihilation. Despite the fact there is now an abundance of evidence, there is still a resistance to the idea of an unseen underworld of antimatter beyond matter because of consensus opinion. As Thomas Kuhn said:

"Truth has as much to do with the consensus of scientists as to the outcome of experiments". [7]

In my lecture at the Royal Institution on 15 January 15 1975, I predicted that galaxies of matter accelerate toward the largest sphere of space due to the electric attraction acting between them and antimatter through the greatest extremity of space. I also predicted that eventually they would meet and annihilate and I proposed that quasars were evidence of this. In that lecture I also predicted that in the apparent expansion of the Universe the furthermost galaxies would be accelerating away from us faster than the more proximate ones. I published this prediction two decades later in 1995 [6] and also suggested that because the acceleration of the galaxies we observe would be toward galaxies of antimatter it would lead to annihilation rather than an indefinite expansion. I suggested that quasars in the outermost reaches of the apparent Universe were evidence of this.

The presence of quasars at either end of the size spectrum of space, occurring in the largest as well as the smallest realms of space, adds credence to the vortex cosmology and increases the credibility of the vortex theory. Occurring in the regions of densest gravitational contraction and greatest universal expansion, they fit the predictions of where matter meets antimatter. However, annihilation of the vortex form of energy would lead not only to the end of matter but also to the end of space and the forces of electric charge, magnetism and gravity associated with matter. If this occurs at both ends of the spectrum of size then there could be a point in the history of time when everything would return to light.

First, exhausted stars would collapse in on themselves to form neutron stars, dwarf stars or blackholes. Then supermassive blackholes at the centres of galaxies could engulf millions of exhausted stars in their gravitational maelstrom. Our modest galaxy has a moderate supermassive blackhole at its centre housing seventeen million exhausted stars in a sphere that would occupy the space of our Sun out to the orbit of Mercury. Many event horizons are much more massive and engulf hundreds of millions of dead stars. These blackholes appear to be stellar crematoriums where the mass of star carcasses are burnt to release heat and light at the blackhole event horizons.

According to the vortex cosmology, when a blackhole forms in matter, an antimatter blackhole also forms in antimatter through the smallest realms of space at their connected singularity point centres. As stars of matter in the blackhole are centralised into singularity, they would be converging on stars of antimatter in the mirror blackhole, which would also be converging into its zero-space point so annihilation between matter and antimatter would take place. With annihilation, the mass of both blackholes would decrease and their gravity would diminish so their survival would depend on assimilating more dead and dying stars in matter and in antimatter. Due to annihilation, the energy released by both blackholes would be continually increasing until it overcomes the grip of gravity, then it could escape and radiate away. Stephen Hawking determined that blackholes leak energy, which comes from matter-antimatter annihilation. [8] If annihilation destroys space, time, mass and gravity as blackholes transform into quasars, then blackholes may be only temporary astronomical formations, just steps in the transition of stars into quasars; of vortex energy into wave energy; of matter into light.

If blackholes undergo transition into quasars then, according to the vortex cosmology, many blackholes should have a quasar associated with them. In fact, new quasars are being discovered with blackholes at their core that are emitting super-fast streams of electrons and gamma ray bursts perpendicular to the blackhole centre. The bursts appear suddenly and then die away. The energy and behaviour of these emissions may be perplexing to astrophysicists, [3] but they fit the cosmology in the vortex theory. The periodicity of these emissions need not be perplexing if they are seen to come from a tussle between the gravity causing particles to conglomerate and the energy of annihilation tending to blow them apart. This is how a gravity geyser would be expected to perform.

The gravity geyser action might occur as a one-off event. Big stars subject to greater gravity that the Sun, burn hot and fast until they are exhausted and collapse under an immense force of gravity provided by their prestigious mass. In the collapse toward singularity matter-antimatter annihilation could occur generating 200x the energy of the burn of nuclear fusion. Caught in the grip of gravity in the collapsing star, the energy would be trapped until it could overcome the gravity and escape. The combination of gravity, collapsing space and enormous energy generation due to annihilation would set up a situation more like an overheated boiler than a geyser. Under immense pressure and phenomenal temperatures high-energy nuclear collisions could occur to generate heavy elements and it would be only a matter of time before an immense explosion would occur, spewing the elements out into space for the formation of mineral-rich planets like our own. In astronomy this is called a super nova explosion. The occurrence of matter antimatter annihilation prior to a super nova explosion is indicated by a burst of gamma rays.

Gamma rays are the hallmark of matter-antimatter annihilation and gamma ray bursts confirm that annihilation of mass occurs through the intense gravitational cores of collapsing massive stars and blackholes. I like to think of these cosmic formations not as astral graves but more like blackhole pupa in which stellar caterpillars morph into quasar butterflies.

In the vortex theory gravitational energy is seen as a product of annihilation between matter and antimatter. That fits with Einstein's belief that gravitational energy is derived from the destruction of mass. When a body falls due to gravity it releases energy on impact. This energy is derived from the destruction of mass because a moving body has a greater mass than a body at rest. As a body falls in the world of matter a mirror image body of antimatter would also be falling, so both would make their impact and release equivalent energy simultaneously. There would be an equal minute annihilation of mass in the matter and antimatter as both bodies come to rest. At rest they would both be closer to their ultimate, complete annihilation. In order to lift the bodies, work has to be done by someone in matter and, simultaneously, by his or her mirror image in antimatter. That would require an equal input of energy to that released when they came to rest.

The fall of every bit of matter on Earth is another step in the inexorable progress toward the ultimate annihilation of mass in the Universe. As you pick up an object that has fallen, you use up energy from the Sun that was acting to keep the particles in the Sun from falling together due to gravity. Eventually, when the Sun's reserves of fuel are exhausted there will be nothing left to keep you alive, let alone lift things! The inertia of gravity will take over and all the matter within our local star will collapse toward antimatter and annihilation. Even though the Sun is too small to form a blackhole, it will first become a red giant before eventually settling down as a white dwarf. Ultimately, it will be swallowed by the supermassive blackhole at the centre of our galaxy.

Annihilation is the ultimate transformation of matter into light. However, if with the annihilation of mass comes the demise of space and gravity, long before all the mass of the Universe is annihilated, before every vortex particle in matter and antimatter unzips, gravity would be declining, and the energy of annihilation would be compressed in collapsing space. It would be forced to race in an ever increasingly tight spiral space path. This would be as a powder keg set for explosion. In the conglomeration of collapsing galaxies, as their mass and energy is compressed toward singularity into ever smaller space, a Big Bang explosion would be inevitable. The death of the Universe would lead to its rebirth. The scene would then be set for a new cycle of expansion of the Universe to begin.

The Big Bang might not be a one-off event in the history of the Universe. If a Big Bang happened in the past, maybe another Big Bang will occur in the future. Maybe the Big Bang was not the beginning of the Universe. Maybe it signalled a rebirth of the Universe. Sages in India speak of the comings and goings of the Universe as the in-breath and out-breath of Brahma. Brahma, Vishnu and Shiva represent the birth, life and death of the Universe but Shiva, the god of destruction, is also the god of rebirth and recreation. There is no end in the Hindu cosmology, only an endless cycle from birth through life to death and then rebirth again. We see this happening everywhere, where apparent beginnings and endings turn out to be part of a cycle. This may apply to the Universe as a whole. If energy is neither created nor destroyed, maybe the Universe of energy was never created nor will it ever be destroyed. Energy exists. All we observe are its unending transformations.

1. Greenstein J. & Schmidt M., *'The Quasi-Stellar Radio Sources',* Astrophysical Journal 140, 1964
2. Fishman C.J. and Meegan, C.A., *'Gamma-Ray Bursts',* Ann. Rev. Astronomy & Astrophysics 33, 1995
3. Tiller W. *Science and Human Transformation*, Pavior, 1997
4. Gehrels et al., *'The Brightest Explosions in the Universe',* Scientific American Dec. 2002
5. Hawking Stephen, *Blackholes and Baby Universes*, Bantam 1993
6. Ash D., *The New Science of the Spirit, College of Psychic Studies,* 1995
7. Kuhn T., *The Structure of Scientific Revolutions,* University of Chicago Press, 1962.

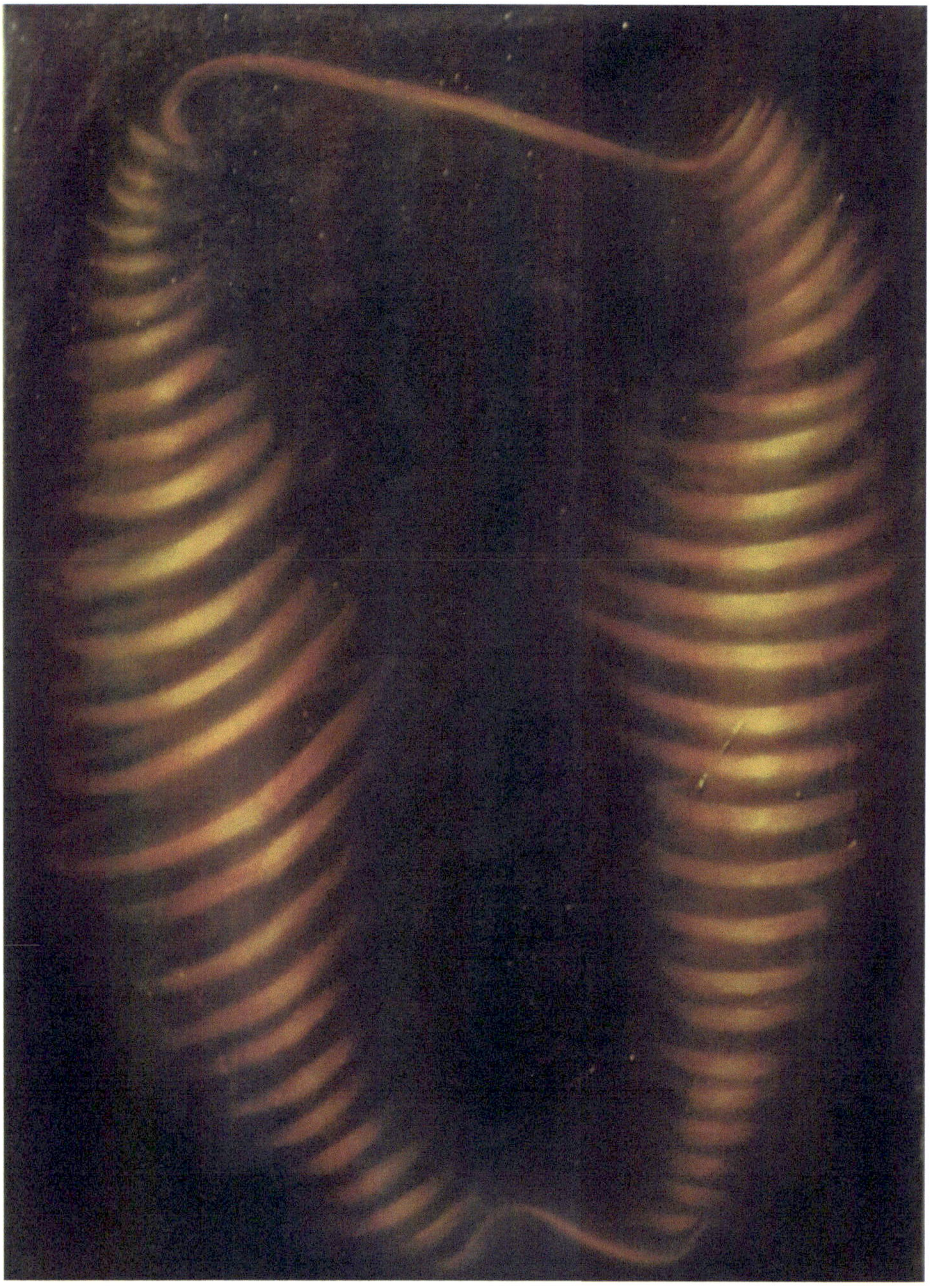

11. Twins - Shane Guffogg

'Twins' illustrate the mirror symmetry of matter and antimatter.

ESSAY 10

Time for Sanity in Science

"There is motion but there are, ultimately, no moving objects; there is activity, but there are no actors; there are no dancers, there is only the dance."

Fritjof Capra

Précis: Our experience of time is influenced by our size. If we were as big as the Universe our days would span billions of terrestrial years. This suggests there may be an element of truth in the story of creation occurring in a week. Many people today dismiss religious stories and believe in science instead, but science can be based on stories too. Einstein said the axioms of a scientific theory cannot be drawn from experiment they must be freely invented. Scientific theories based on invented axioms are stories. In his book, *Sapiens,* Yuval Noah Harari listed large-scale human cooperations as stories but he didn't include scientific theories because in his book, *21 Lessons*, he revealed the scientific theory he believes in is materialism, which is a story. Sapiens are quick to question the beliefs of others but are not so quick to question their own beliefs.

Most scientists and academics believe in the material hypothesis which is a story. Quantum mechanics is an incomprehensible story based on Heisenberg's principle but if its base were shifted to accommodate the vortex of energy it could come into alignment with common sense and become straightforward, self-evident and comprehensible for the non-scientist as well as the scientist because the vortex theory is not based on a story. Its axiom, that the smallest particles of matter are vortices of energy, was not invented. It came from a direct observation of the quantum world in the pre-scientific era when yogis used techniques of mind over matter to probe the atom. This yogic power may have been lost in the modern era but its potency was proven by the way it enabled yogis to anticipate Einstein's discovery of the equivalence of mass and energy.

In summary, with the advent of the vortex theory there is no need to depend on stories in science. There is no need for inventions like quarks to explain the foundations of matter or virtual particles to explain forces. The theory of the vortex of energy provides a realistic appreciation of forces and a new foundation for quantum mechanics, which is clear and simple to understand. There is no need for Higg's boson either as spin in the ball vortex of energy can account for mass. Electrons and proton vortices are real particles of energy that carry the forces of nature and the new version of quantum mechanics, based on the vortex theory, is seamless with Newtonian mechanics. It makes sense of the world we live in.

Common sense is brought into the transitions between mass and energy. They are seen to be just changes of form between vortex and wave. In these changes the temporary swirls of energy generated in high-energy laboratories, which appear as short-lived particles, are not real particles. They possess mass for microseconds because they are wave forms of energy pressed through natural vortices that have the vortex form imposed upon them. They revert to their natural wave form as soon as they are released from the press. Physicists have become obsessed with these momentary transitions between vortex and wave. They have built massive machines to study them. The importance placed on these transitional states of mass may be unwarranted. Standard physics may make sense to physicists but it makes no sense to most of us. It excludes the majority from the debate about the structure of the Universe and the meaning of life. To many of us science has become insane.

The vortex theory is an attempt to bring sanity back into science. In place of maddening complexity and incomprehensible maths, there is refreshing simplicity. It fulfils the promise at the end of Stephen Hawking's book A Brief History of Time, that if ever a complete theory is discovered it should be understandable in broad principle by everyone, not just a few scientists.

Fig 91 Albert Einstein

~

Essay: In his 1905 theory of relativity, Einstein revolutionised our understanding of space and time. In the vortex theory there appears to be a self-evident relationship between time and size of space. While the thirteen and a half billion years since the Big Bang may appear for us to be an immense span, in the timescale of the Universe it could be as a breath. This is because the distances of space and periods of time in the Universe only appear large to us because we are miniscule compared to the Universe. If we take our measures of time from our position in the Alician Dimension of size, then intergalactic time would seem large to us because we are little compared to a galaxy. Time is relative to the size of the observer. I came to this realisation when I passed through the students union one day in 1971. As chance would have it a television was showing an Open University film of animals slowed down. I was transfixed by cats moving with the fluidity of panthers and insects flying like birds. Seen normally, their movements are fast and jerky. It struck me this might be the way they appeared to us only because we are bigger than them. If I were the size of a mouse, a cat might come at me like a panther and butterflies wing as birds. This is because my experience of time is relative to my position in the size spectrum of space.

It was a sunny summer afternoon as I walked out of the college and into Holland Park to think about what I had just seen. I did what Einstein did and ran a thought experiment. I imagined a space giant who eyed the Earth as a football then decided to use it as a clock instead. He counted each rotation of the Earth as a second and the orbit of the moon as a minute and the annual circuit of the Earth round the Sun marked out an hour for him. His year was our century and we were as bacteria on the football to him. His lifespan would be millennia to us. If the giant was as big as the Universe, his day would be a billion years to us. On that scale the Universe would have been around for a couple of weeks and the appearance of the Earth and evolution of life upon it would have happened in a matter of days. As an up-and- coming Bachelor of Science pondering in the park that day, I became a little more humble about the six days of creation...and on the seventh day He rested. I needed a rest so I sat on a bench and watched the peacocks. Were they made so beautiful in a day or did they evolve that way over millions of years?. That was for creationists and evolutionists to debate. The best I could do was to reconcile their time scales. Maybe they were viewing the same story from different perspectives of time. I saw great potential in the vortex theory to bring unity and reconciliation to our understandings of the Universe.

Half a century later I wrote the first three of these essays in six days in my determination to tell the world that Einstein was right and Heisenberg was wrong. But as I rested on the seventh day, I realised that while Einstein was right to challenge the uncertainty principle, from hindsight I saw it had played an essential part not only in the evolution of quantum theory but also in my own journey. The uncertainty principle may have failed the neutron test and the vortex theory may offer an alternative approach to quantum mechanics, but without Heisenberg's tenacity in the face of the discovery of the neutron there would be no quantum mechanics to reconsider and I doubt if I would have developed the vortex theory. This is because Yukawa's prediction of mesons was based on Heisenberg's formula and that stimulated Cecil Powell's cosmic-ray research, which kick started high-energy research into quantum mechanics on one hand and my vortex theory on the other. I would not be where I am now without Heisenberg's uncertainty principle. For me this was an important lesson in seeing the value in every endeavour which has brought us to where we are now.

Quantum mechanics and the uncertainty principle have proved to be immensely useful and for most people 'useful' is more important than 'truthful'. It is unlikely we would have the high tech computer age without the uncertainty principle and the quantum mechanics that stemmed from it. Professional physicists have got used to quantum mechanics and they don't need to understand it if the maths for it works. Nonetheless, for the sake of the philosophy of science, a sound theoretical underpin for quantum mechanics is needed to replace the uncertainty principle because it did fail the neutron test. A visual description of quantum mechanics that is simple and self-evident, easy to comprehend and in compliance with common sense, would be helpful for the popular understanding of quantum physics.

The uncertainty principle was a brilliant insight but quantum mechanics can survive without it. Ultimately it is just a theory and it wasn't even invented by Heisenberg. According to Arthur Miller in *137*,[1] Heisenberg and Pauli often exchanged equations to work on and when Pauli was tackling a mathematical problem for Heisenberg he discovered it was impossible to determine with accuracy both the momentum and position of a subatomic particle. He wrote to Heisenberg about this and Heisenberg was delighted with Pauli's discovery. He developed it into his own theory of the uncertainty principle and accepted the 1932 Nobel Prize for 'his discovery of quantum mechanics'.

Science is built on theories and Einstein said that the axiomatic basis of a scientific theory cannot be drawn from experiment; it must be freely invented. But the axiom of the vortex theory was not invented. It was not based on speculation. The axiom of the vortex theory, that the smallest particles of matter are vortices of energy, was discovered by yogies in ancient times through a direct observation of the quantum world attained through meditation. What the yogis achieved through mind over matter has proved its worth by anticipating Einstein in his discovery of the equivalence of mass and energy and by providing a way to visualise quantum spin.

Quantum physics is driven by highly dedicated people. It is rich and full of personal adventure and remarkable discoveries but in its complexity it has lost touch with reality. Physics should advance philosophy as well as technology and it should be truthful as well as useful. The endeavour to establish truth in physics is important because with improved education ever more people are looking to science instead of religion for the truth. Some people are seeking a meeting point between science and spirituality. As the vortex theory is rooted in Yoga it offers a strong possibility of achieving that. With its strong root in Yoga the vortex theory has the potential to dissect fact from fable in both religion and science.

Fig 92 Yuval Noah Harari

In his book *Sapiens: A Brief History of Humankind,*[2] Yuval Noah Harari wrote:

> "Any large-scale human cooperation – whether a modern state, a mediaeval church, an ancient city or an archaic tribe – is rooted in common myths that exist only in people's collective imaginations...Yet none of these things exists outside the stories that people invent and tell one another. There are no gods in the Universe, no nations, no money, no human rights, no laws and no justice outside the common imagination of human beings."

In his list of stories Harari didn't mention science. He is an academic and like most academics he has faith in science. It is obvious he believes that science is based on fact not fiction because if he thought science was full of stories he would have said so. However, science is full of stories masquerading as science and the main one is materialism and Harari believes in that story. This is evident from this sentence from his book, *21 Lessons* [3]:

> "In itself the Universe is only a meaningless hodgepodge of atoms."

That is a pithy summation of the pessimistic philosophy of materialism. Most academics, scientists and philosophers won't admit to materialism being a story but it is. Materialism was disproved by Einstein in his equation $E=mc^2$, which revealed mass is not formed of material substance, it is formed of energy. Materialism is fiction not fact. It is no longer scientific. Materialism is a story that most people believe but consensus belief does not turn fiction into fact and people don't like to be told that their beliefs are stories. They prefer to tell other people that their beliefs are stories. This is a common human trait. Sapiens are quick to question the beliefs of others but are not so quick to question their own beliefs.

To appreciate how unscientific the story of materialism is, imagine rain falling to form streams tumbling down mountain sides in the Alps. These join the torrent of rivers that pass through hydroelectric plants where the fall of the water is converted into the spin of turbines and then the flow of electricity. The electricity is then fed into CERN where it is used to accelerate protons. As the protons collide, in the intersecting rings of the Large Hadron Particle Collider, their arrested motion is transformed into mass. The kinetic energy derived from the motion of falling water has been converted into new particles of matter. In the fall of rain and the tumble of streams, the torrent of rivers and the spin of turbines, the flow of electricity and acceleration of protons no material substance was transferred, clickety-clunk, through each transition to form the new particles. Activity alone was transferred between each step of the process. As nothing went into the newly formed particles of matter but activity, obviously, they can be nothing but forms of activity; the activity we call energy. The vortex theory is just an attempt to reveal in a clear and simple way how activity can form particles of matter, albeit transient, and how that ties together with the rest of physics.

The creed of materialism, also called physicalism, is that seeing is believing, and materialists accuse idealistic people, not avowed to this point of view, of being naïve. Now the tables have turned and it is the philosophy of materialism that has proved to be naïve but some people are slow on the uptake. I met a philosopher once who dismissed me very rudely when it became clear to him that my theories were more inclined toward the philosophy of idealism than materialism.

Fig 93 Dartington Hall

It was a glorious summer afternoon in 1979 when, with my family, I arrived for tea in the garden of the philosopher in residence at Dartington Hall. He heard I had a theory with philosophical implications and gathered a small group to hear me out.

We had driven to Dartington in Devon from Millbrook in Cornwall and were parched. Anna, my wife, and our four little children were hungrily tucking into sandwiches, scones and tea when I opened my big mouth. Had I stuffed it with a scone we might have stayed on to enjoy the tea party. Instead, unable to contain my impatience to hold forth, I let out that Einstein's equation $E=mc^2$ implied the fabric of the Universe was pure movement and therefore the Universe was more in the nature of mind than material reality. The philosopher first froze then became incensed. He spluttered something about Bishop Berkley and shouted he would not entertain someone who preached idealism. He shooed us out of his garden like geese. We had to leave our tea undrunk, drop our scones and scarper. The children were upset. Anna couldn't believe it. All I could do was laugh. I had met this attitude of intransigence in the name of science many times before. One of the reasons I felt an affinity for Einstein is because he was thrown out of the tea party of theoretical physics because of his philosophy. I have admiration for the stand he took against Heisenberg and quantum mechanics because science needed a shakeup. Science was set in the story of materialism and the scientific community were resistant to change it. Einstein stood for realism but his famous equation lent support for idealism and quantum mechanics has encouraged belief that consciousness is the bedrock of reality which is central to idealism.

Over the years my work has been put-down repeatedly by people who embrace quantum mechanics. When my father sent a copy of my first paper to the then Astronomer Royal, Sir Martin Ryle, who had been a boyhood friend of his, he replied that we don't need the vortex theory because quantum mechanics provides adequate answers. Quantum mechanics may satisfy professional physicists and astronomers but it is out of reach to the lay public, and even physicists don't appear to understand it.

To quote Richard Feynman:

"I think I can safely say that nobody understands quantum mechanics". [4]

My endeavour has not been to put down quantum mechanics but to make it more understandable. Scientists persevered with Heisenberg's quantum mechanics even though they didn't understand it because it worked well mathematically and served the needs of technology. But in doing so they overlooked our philosophical needs. Quantum mechanics may tell us how things work but not what things are. Mostly we aren't interested in equations and technical jargon. We leave that to the professionals. Most of us want to understand things in plain language. We have a right to that and even Heisenberg admitted professionals require it too when he wrote:

"Even for the physicist the description in plain language will be a criterion of the degree of understanding that has been reached." [5]

Gary Zukav spoke of the value of a non-mathematical approach to physics when he wrote:

"The fact is that physics is not mathematics... Stripped of mathematics, physics becomes pure enchantment." [6]

I have not endeavoured to prove my theory with complex mathematical equations. My aim has been to make the physics based on those equations more understandable. My greatest joy is to present physics in plain language, to make it more appealing and reveal its enchantment. My vortex theory is a personal endeavour to see if it possible to approach quantum mechanics through art. I have used pictures, simple descriptions and analogies to get to grips with quantum physics and I now want to share my experience with you.

I have used the visual model of a ball vortex to replace the classical billiard ball model for the material corpuscular particle. I perceive a real need for this because with the abandonment of realism in quantum physics we have been left without a way to visualise quantum mechanics.

Physicists have disappeared in a cloud of complexity leaving us with the classical model of the material particle. The vortex model from Victorian times has been lost. The vortex atom was an improvement on the material view of the atom. The fact that it failed to explain new discoveries about the atom was only because it turned out that the atom was not the most fundamental particle of matter. Applied to subatomic particles the vortex is amazing. It explains practically everything. But no one thought to go subatomic with the vortex theory. The vortex particle was ditched and the material particle rolled on.

Fig 94 Richard Feynman

At its core the vortex atom theory presented matter as a form of motion, and this view was supported by Einstein when he equated mass with energy. Had the vortex been retained it would have provided a self-evident model for how energy forms mass. Instead, scientists, philosophers and those who call themselves Sceptics, continue to promote the myth of materialism as though it was true despite the fact it has been disproved by Einstein's equation $E=mc^2$. While Sceptics say they believe in nothing, most believe that science is still based on the atomic hypothesis, which is the foundation of materialism. This was summarised aptly by Richard Feynman:

> "If in some cataclysm, all of scientific knowledge were to be destroyed, and only one sentence passed on to the next generation of creatures, what statement would contain the most information in the fewest words? I believe it is the atomic hypothesis... that all things are made of atoms – little particles that move around in perpetual motion." [7]

If everything is made of energy all things would be made, not of little particles that move around in perpetual motion, but little particles of perpetual motion. Most physicists don't get this because while they are satisfied that everything is formed of energy they are not clear about what energy is. To quote Richard Feynman again:

> "It is important to understand that in physics today we have no idea what energy is." [8]

Energy is a measure of activity. That is how it is defined in physics. Einstein's founding contribution to quantum theory was to show that energy is particulate. So, energy consists of particles of pure activity. Because energy has no mass and no substance it cannot be assumed that there is anything underlying the activity called energy that makes up everything. Most scientists and philosophers may balk at the idea that the Universe is based on activity where nothing exists that acts. They may despise the philosophy of idealism, that reality is primarily mental. But enlightened quantum physicists have come to terms with the unavoidable conclusion that particles of energy are more in the nature of thoughts than things. Physicists, philosophers and Sceptics, embedded in the philosophy of materialism may throw up their hands in horror and turn idealists out of their gardens but flat earth thinkers did the same to people who said the Earth was round. Realising the Earth is round didn't change people's lives it just helped them to be more realistic about the world they were living in. Likewise the vortex theory may be idealistic in its core philosophy but it embraces realism by providing a realistic image of quantum reality. The issue I have is with materialism or physicalism as it is now called. This philosophy is not true, it is just a story.

In the philosophy of physicalism a particle or a field is presumed to exist and then move. In materialism, mass comes first and energy is merely a measure of its movement, so that comes after. But Einstein's equation $E=mc^2$ makes it clear that energy comes first and mass comes after. The problem for physicists is to understand how energy as pure motion, which is no thing, can form mass which appears to be something. This issue can be resolved by treating a massive particle as a particle of vortex spin. Vortex spin is a powerful tool to explain the greatest enigma in physics: how energy forms mass. Suggesting

Fig 95 Democritus

that energy spins in ball vortices to set up massive particles not only resolves the enigma but from that resolution flows a clear and simple account for practically everything else in physics. In the process, classical materialism is dismissed as delusion and the famous statement of its founder, Democritus, that only atoms moving in space are real; everything else is opinion, is just his opinion.

Quarks are an example of the determination of materialists to find the ultimate particles of material substance but quarks don't exist. After decades of searching, no one has ever seen a quark in the free state. The quarks scientists claim to have detected are an interpretation of experimental data and the interpretations of experiments are liable to be biased by the deluded state of seeing what is believed.

When physicists bombarded protons with high-energy electrons at the Stanford Linear Accelerator in California, the experiments suggested that protons contain swarms of particles. Because physicists were looking for quarks they assumed the swarms were quarks. However, in the vortex theory the particles swarming in protons are considered to be captured energy swirling inside them. Quark theorists also predicted a particle containing three strange quarks. They were jubilant when the Omega minus particle was discovered to decay in three steps, which they took to be its breakdown into three strange quarks. But the vortex theory offers an alternative account for the Omega minus with less arbitrary elements than the quark theory.

All experiments are subject to interpretation and, despite Karl Popper's protestations, science is not in the business of disproving theories. Most scientists are inclined to interpret experiments in favour of theories they want to prove. It is only natural for people to seek success not failure. Seeing is believing, and physicists are inclined to see what they believe in the interpretations of their experiments. Scientists may say they have performed an experiment to prove a point, but we cannot trust

Fig 96 Carl Popper

their interpretation of the experiment as it may be biased by the point they want to prove. The interpretations of scientific experiments cannot be certain if they are biased by the beliefs or inclinations of those who interpret them. Scientists may deny any bias in their interpretation of an experiment but we cannot always trust what people say, which introduces another level of uncertainty into science. The truth is that everyone is subject to bias, not only from beliefs but from funding, upbringing, education and culture. Experiments may be true but the interpretations of them may not be. This is just one of the weaknesses in the scientific method which introduce uncertainty into physics.

The philosophy of science is to test our beliefs and assumptions against the evidence of experiments. This is commonly expressed as everything should be evidence based. However, as well as the unreliability of experimental interpretations by scientists, we are faced with their tendency to cherry pick experiments and consider only those providing evidence they are willing to accept. It is not uncommon for scientists to ignore experiments or discount discoveries that are in conflict with their beliefs or the consensus view of reality. The discovery of the neutron was an outstanding example of this.

Another weakness in the scientific method is the reliance on predictions. Throughout history humans have tested oracles and prophets by their ability to predict future events and this expectation has been passed onto the sages of science. During a solar eclipse in 1919, the Royal Society tested a prediction made by Einstein in his general theory of relativity. The result of its success was spectacular. Overnight Einstein became the most celebrated scientist in the world and his prediction that mass distorts spacetime was considered proven. But Einstein's sensational success didn't prove his prediction;, rather, it revealed a weakness in relying on predictions for the success of a theory in science. In the vortex theory there is a subsequent contradictory interpretation of the eclipse experiment, which suggests that space is curved not because it is distorted by mass but because it is an extension of mass. Einstein's assumption was arbitrary whereas the vortex account for space curvature is based on a single assumption that has a wider base in physics and unifies relativity and quantum mechanics. On this count, Occam's razor can be invoked in favour of the vortex theory against the general theory of relativity. This highlights the weakness in putting too much store by predictions.

Another example of this weakness is Yukawa's prediction of mesons based on Heisenberg's formula. Despite the contradictory evidence of the neutron, the success of this prediction when it was tested in cosmic-ray research was taken as proof that the principle of uncertainty was true. However, the vortex theory has shown an alternative explanation for Yukawa's prediction that meson energy exchanged between protons can account for the strong nuclear force so again Occam's razor can be invoked. If both Einstein's and Heisenberg's theories are displaced by the vortex theory then the prediction criterion of the scientific method would have to be reviewed, or even revoked, especially as the vortex theory offers an alternative to Einstein's theory for gravity that unifies gravity and the electric force. If the scientific method is thrown into doubt then the scientific criteria for truth can no longer be trusted.

In his book *Hidden Journey*, Andrew Harvey quoted an Oxford academic as saying:

> "Only scientific criteria for truth are valuable and mystics are pathological cases." [9]

Mystics in ancient India used their supernormal powers to discover that vortex motion is the way energy forms mass. Scientists used their normal powers to utilise Einstein's discovery of the equivalence of mass and energy to develop weapons of mass destruction. Who, I ask you, have demonstrated evidence of pathology – mystics or scientists? Whose criteria for truth are proving more valuable:, that of yogi mystics, which led to the vortex theory or that of the scientists, which led to the general theory of relativity and quarks and Higg's boson and virtual particles that appear and disappear within the bounds of uncertainty?

Quantum mechanics based on the uncertainty principle has been highly acclaimed because predictions of virtual force-carrying particles, formed of borrowed energy, appear to have been vindicated by experiments. But the alternative theory for vortex force-carrying particles, formed of real energy, satisfy the same experiments. The action of infinite extending dynamic vortices of energy is a more plausible account for forces than the presumption that subatomic particles exchange virtual force-carrying particles, within the constraints of Heisenberg's uncertainty formula, with precision, over infinite distances. Gauge theories, in which virtual particles are presumed to feel the presence of other virtual particles is a clear case of speculation heaped on speculation. Uncertainty about the uncertainty principle throws doubt on the theories of virtual particles and gauge theories even without considering Occam's razor. There are so many arbitrary elements in the current theories of quantum mechanics that none of them would survive the test of Occam's razor against the vortex theory with its single axiom that explains practically everything.

Fig 97 The falcon analogy

In the pleasant plaza of quantum physics a peregrine has arrived among the pigeons. Not only are the accepted theories in quantum mechanics being picked off one by one but the scientific method itself, and indeed the scientific criteria for truth, are up in the air.

The success of quark theory is largely due to the vindications of predictions in the theory. The ability of the vortex theory to provide alternative interpretations of experiments, thought to prove quark theory, undermines the credibility of quark theory and further weakens the prediction criterion of the scientific method. The falcon is onto quarks.

With heaped speculations the quark theory is unlikely to survive the test of Occam's razor against the vortex theory. If quarks were discovered in the free state they would be more plausible but their existence is entirely reliant on the interpretations of experiments. To survive, quarks must be detected repeatedly in the free state

The same applies to Higg's account for mass. It depends on the arbitrary assumption of the Higg's field permeating space. That is ominously reminiscent of the speculation in classical physics of a field called the luminous ether, which has been disproved. Then there is the additional assumption that Higg's bosons are generated by excitations in the hypothetical field. Higg's bosons can't be detected because they decay too rapidly so they never appear in the free state either. They were predicted to leave a trail of mu-mesons or muons when they decay and the rare appearance of four muons in an experiment conducted at CERN in 2012, was hailed as the discovery of a Higg's boson. The overdramatised event was nothing but an observation of debris expected in the decay of a Higg's boson. It could have been a total coincidence that four muons were synthesised under the expected circumstances by the enormous energy developed in the colliders. Also the possibility has to be taken into account that the interpretation of experimental data on that occasion was biased by the need make a dramatic discovery in fundamental physics to justify the eight billion eros spent on building the Large Hadron Collider at CERN? With so many arbitrary elements in the Higg's boson theory it would never survive the test of Occam's razor against the vortex theory even if the evidence was beyond doubt in the field of scientific integrity.

To be real, particles must be apparent and they must be durable. This was made clear to me by Susan Thompson, the writer who kindly set aside her writing to help me edit these essays. She was drawn to think deeply about the neutron because there was something about it she was unsure about. When I asked her what her problem was she replied:

"You call it a particle but how can it be a particle in a permanent state if it only exists temporarily then breaks down into other particles."

Suzie had stated the obvious. A particle cannot be presumed to be real if it decays into other particles because it may be just a transition stage between real particles. As well as neutrons, pions, muons and neutrinos have been detected in the free state and physicists have presumed they are real particles. However, they are not durable in the free state. Pions, muons and neutrinos decay rapidly into particles of radiant energy. Neutrinos are only presumed to be durable. In the vortex theory, free state pions, muons and neutrinos are explained away as operations of the quantum laws of motion. They are treated as transitions between energy in vortex spin and energy in wave vibration. As such they would be momentary transitions between mass and propagating energy. This alternative account for pions, muons and neutrinos throws the standard theory of physics into the air. That is a fat pigeon for the falcon.

Neutrinos are only presumed be durable. There is no evidence that they are. Quarks and Higg's bosons are entirely hypothetical as they are not apparent. Particles generated in the particle accelerators appear for short periods of time but they may not be real particles as they could be just mass-energy transitions. None of these particles can be classified as real particles alongside electrons and protons that are very apparent and survive for exceedingly long periods of time.

Physicists have treated the particles they have generated in their accelerators as real because Heisenberg's uncertainty principle allows for particles to come into existence for short periods of time and the shorter the time the higher the energy and greater the mass they can possess. By coincidence the high mass, short lifespan profile of the high-energy particles allowed by the uncertainty principle, fits the quantum laws of motion; which suggest an immediacy in the reversion of captured mass to radiant energy. The greater the escaping mass the less stable it would be, so the less time it would take for a swirl to unravel and radiate away. This calls into question the interpretations of high-energy physics. It is the interpretation of experimental data, not Heisenberg's principle, that is the real uncertainty in quantum physics. Any ambiguity in the interpretation of experimental data undermines the value of that data so the importance placed on particles generated in high-energy physics may be unfounded. Particles that are apparent in high-energy experiments cannot be assumed to have a place in nature if there is a chance they could be just anomalies of the high-energy experiments. Unless they are also observed in cosmic events they cannot be presumed to be natural. High-energy facilities may be just fabricating swirls of energy that exist momentarily at given quantum states. There is a need for clarity in physics for what are the real natural particles, what are apparent particles appearing in momentary mass-energy transitions and what are virtual particles that exist only as mathematical constructs serving theories, such as Richard Feynman's virtual photons.

The issue of justification must also be taken into account. Dramatic discoveries are needed to justify the expense of construction and running high energy laboratories such as CERN, Fermilab and SLAC. The alleged discoveries of quarks and Higg's boson may have been interpretations of experimental data biased by needs to justify not just the initial cost of the facilities but the ongoing cost of the army of scientists and technicians employed in running them. And what is the value of the experiments conducted in these high energy labs, if scientists are not discovering real particles. Is the expense justified if the research is into particles that can only be described as phantoms?

As far as Susan Thompson is concerned, particles are real if they are detectable in the free state and have the durability of the naturally occurring particles such as protons, electrons and photons. I call this 'The Thompson Principle' because it was Susan Thompson who first spoke to me of the need for a defining principle for the classification of particles. She is not the only clear-thinking lady to realise the need for a clean sweep in particle physics. The astrophysicist Sabine Hossesnfelder wrote an article in The Guardian, on September 26, 2022, headlined: 'NO ONE IN PHYSICS DARES SAY SO, BUT THE RACE TO

INVENT NEW PARTICLES IS POINTLESS' Hossesnfelder laid bare the danger of unbridled speculation in particle physics. The vortex theory offers a clear and simple way out of this mess. It offers a way of understanding particles in line with the edict of Ernest, Lord Rutherford, the father of nuclear physics, that these fundamental things have got to be simple.

In summary, the vortex theory provides a new approach to quantum mechanics based on a clear and simple account of electrons and protons as real particles formed of vortex energy. Photons are treated as real particles of energy without mass because they are not based on vortex spin. They are wave forms of energy. The wave and vortex particles of energy can interact and interchange and, as they do so, short-lived massive particles may appear as transitional vortices of energy. In the vortex approach to quantum mechanics, electrons and protons are treated as permanent forms of vortex energy that account for mass, space and the electric and magnetic forces as well as gravity. Pions are treated as impermanent forms of vortex energy in the free state but inside protons they are permanent. There they contribute to proton mass and account for the nuclear binding force in line with Yukawa's's prediction. Electrons in the atomic nucleus support this binding by contributing negative charges to reduce the level of positive charge repulsion between the protons. The unstable particles generated in high-energy laboratories possess mass, but they are not considered to be real particles in the quantum vortex mechanics. They are transient whirlpools of energy which possess mass only while they are pressed into the vortex form. These transitional forms of mass immediately revert to the wave form as soon as they are released from the press of real vortex particles.

The mass-energy transitional states observed in high-energy laboratories may be of interest to scientists, but they have no meaning or purpose in the daily lives of those who have to carry the tax burden of paying for the construction and running costs of the facilities.

The fact that the majority of physicists believe in the standard theory of physics does not mean it is the correct description of quantum reality. Consensus opinion does not prove a theory. The majority of scientists believe in quarks and the virtual particles of quantum mechanics because that is what they have been taught but this weight of opinion has not proved these theories. The time-tested criterion for deciding on theories is Occam's razor and that is where the vortex theory scores. It provides an account for quantum physics based on the single assumption that particles of energy occur as vortices as well as waves. By introducing spin alongside waves as a primary form of energy it unfolds as a complete theory offering a new way of understanding quantum mechanics that is readily available to everyone. The vortex theory has been developed over a sixty year period and limited personal funding has helped to maintain its integrity and impartiality.

Stephen Hawking concluded his book, *A Brief History of Time* [10] with the paragraph:

"If we do discover a complete theory, it should in time be understandable in broad principle by everyone, not just a few scientists. Then we should all, philosophers, scientists, and just ordinary people be able to take part in the discussion of the question of why it is that we and the universe exist..."

What is the likelihood of a complete theory being discovered that is also understandable in broad principle by everyone, not just a few scientists. Physicists invariably develop theories with mathematics. In order to be understood by everyone the complete theory could not be couched in math. There is a saying that for every equation in a book there is a loss of 75% of the readers. The only exception is $E=mc^2$. A complete theory that enables philosophers, scientists, and just ordinary people to be able to take part in the discussion of the question of why it is that we and the Universe exist, would have to be developed and presented in a way that ordinary people can understand.

Neils Bohr succeeded because his iconic model was visually appealing but Heisenberg and Pauli, who came in his wake to develop quantum mechanics, were unable to find simple models to depict their discoveries. Their inability to provide realistic visual images of quantum reality, especially quantum spin, led them to despair of realism. They accepted spin but vortex spin would have been an unacceptable step back for them because their mentor Bohr had explained the atom in a way that swept aside the vortex atom theory. But a century has passed and with hindsight we can see how effective the vortex model would have been had it been applied to subatomic particles instead of the atom and to energy instead of the ether.

The image of the subatomic particle as a vortex of energy has visual appeal. It is art and the journey I have been on has been to show how effective art can be in unravelling the mysteries of the quantum world. Visualisation is necessary to make sense of reality and people don't like uncertainty. They rely on certainty to live their lives. In these essays I have endeavoured to unmask the uncertainty principle and show the value of yogic philosophy in providing us with the realistic image of the vortex of energy, which can help us visualise how energy forms mass and fully appreciate Einstein's remarkable discovery.

I have always had implacable faith in Albert Einstein. He was certain he could lead us to a theory for everything that is suitable for everyone. In 1905, when he was just a clerk in the Swiss Patent Office, he proved the atomic theory, proposed the theory of relativity, established quantum theory, and ascertained that mass is equivalent to energy. He was well on the way to developing a complete theory but, as Walter Isaacson said, in his quest for a unified theory, there seemed a lack of fundamental physical insights guiding him. If Einstein had been aware of the yogic insight he might have made his breakthrough. It was published in January 1905.[11] In these essays I have endeavoured show that if Einstein had stumbled across the yogic perception of vortices of prana, alongside his formula $E=mc^2$, he might have achieved the complete theory of unification in physics and in his contest with Heisenberg he would have been proved right.

1. Miller A. *137: Jung, Pauli, and the Pursuit of a Scientific Obsession,* P.100, W. W. Norton & Company, 2010
2. Harari Y.N., *Sapiens: A Brief History of Humankind,* Vintage, 2014
3. Harari Y.N., *21 Lessons for the 21st. Century,* Jonathan Cape, 2018
4. Feynman R., *The Character of Physical Law,* BBC Publications, 1965
5. Heisenberg W., *Physics and Philosophy,* Harper & Row, 1958
6. Zukav G., *The Dancing Wu Li Masters,* Rider, 1979
7. Feynman R., *The Feynman Lectures on Physics, Vol 1.,* Addison Wesley, 1963
8. Feynman R., *The Feynman Lectures in Physics: Conservation of Energy,* Caltech, 1963
9. Harvey A., *Hidden Journey,* Rider, 1991
10. Hawking S., *A Brief History of Time,* Bantam Press,1988
11. Yogi Ramacharaka, *An Advanced Course in Yogic Philosophy,* The Yogi Publication Society, January 1st 1905

12. An English Silence - Shane Guffogg

For me, this image depicts energy waves swirling inside an extending quantum vortex. - Ash

List of Art and Illustrations

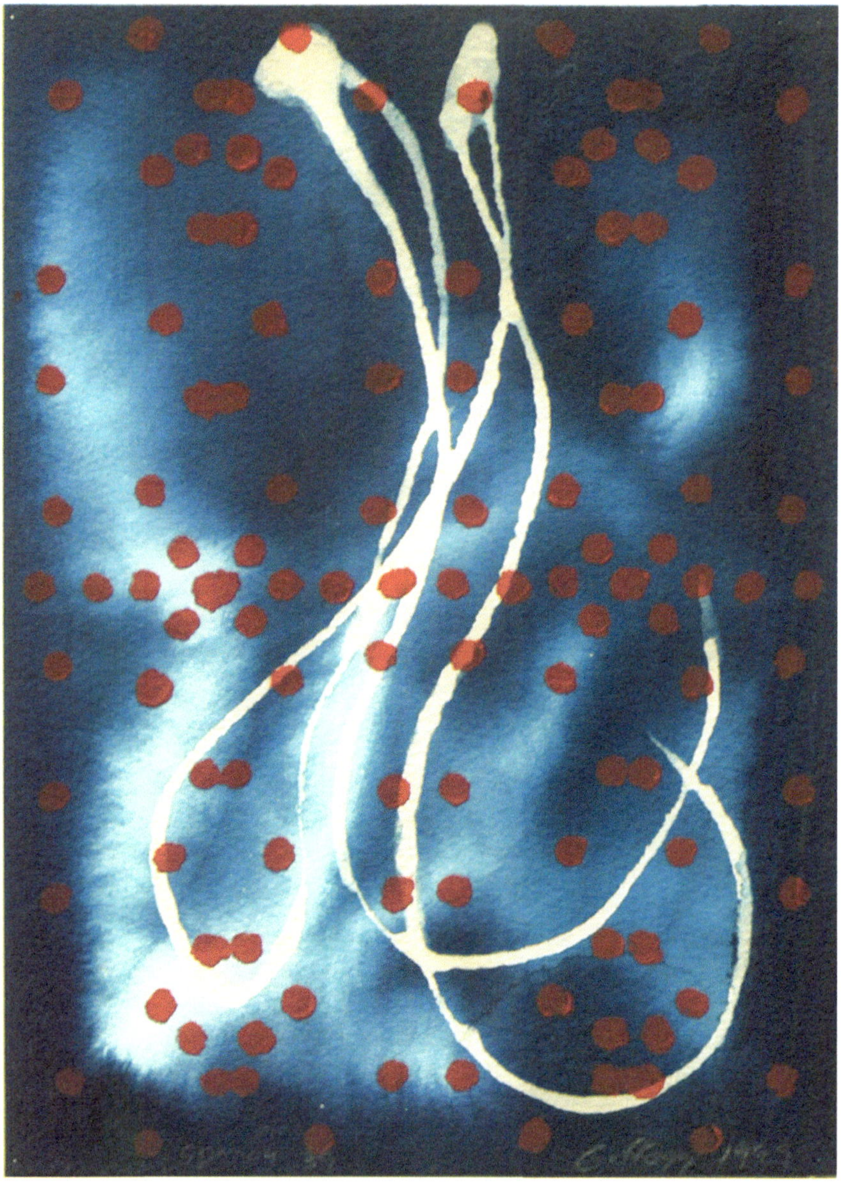

13. Spanda - Shane Guffogg

"*Spanda* is a Hindu word meaning the energy that created the universe. This painting was made as part of my pattern series. In that series I would allow my subconscious to create the initial marks which often resemble ribbons...These paintings are a bridge between my conscious and subconscious. And the patterns are the manifestation of thought" - **Shane Guffogg**

Glossary

Alpha-particles	*are positively charged helium nuclei released by radioactive atoms.*
Apparent particles	*are detectable but not durable*
Atomic theory,	*attributed to Democritus, is that matter is composed of minute indestructible particles called atoms meaning that which cannot be cut.*
Atoms	*are systems of subatomic particles consisting of electrons in defined quantum states orbiting protons and neutrons in a central nucleus. This model for the atom was established by Niels Bohr. The atom is mainly space. If the nucleus were the size of a golf ball the closest orbiting electron would be about two miles away.*
Ball vortex	*a corpuscular or spherical vortex without a measurable axis of spin modelled by a ball of wool.*
Baryon	*a group of heavy subatomic particles including the proton.*
Beta-particles	*are negatively charged high speed electrons released from radioactive atoms.*
Blackholes	*the densest centres of matter with the most powerful gravity that can trap light.*
Bosons	*are subatomic particles defined by Planck's constant, h. In the vortex theory they are considered to be particles formed of two strings of energy; two lines of the motion of light.*
Carbon	*an element with six protons in the nucleus of its atoms.*
Capture theory	*a part of the new vortex theory that describes the ability of vortex particles such as the proton or electron to capture or trap wave particles of energy such as quanta or photons in their vortex structure and transform them completely or in part into vortex motion; into mass. The vortex takes no action in the process of capture. The wave train energy drives into capture by virtue of the path of spiral space presented to it by the vortex.*
CERN	*the European laboratory for nuclear and high energy research near Geneva.*
Chemistry	*the study of atomic interactions.*
Copenhagen Institute of Theoretical Physics	*established in Copenhagen University by Niels Bohr in 1921, originally funded by the Carlsberg brewery and later by the Rockefeller Foundation.*

Copenhagen Interpretation	*a philosophical position taken by the Copenhagen Institute of Theoretical Physics that includes the denunciation of realism as naïve and the contention that quantum reality is said to exist only when observed.*
Electric charge	*a force of attraction or repulsion between primary particles of matter.*
Electric dipole on the neutron	*is the measure of a minute negative and positive polarity on a neutron considered to be too weak to be taken as conclusive.*
Electromagnetism	*electric and magnetic fields.*
Electron	*a light subatomic particle with a negative charge and life span estimated in quadtrillions of years.*
Energy	*a measure of the activity of a system.*
Ether or Aether	*a presumed frictionless medium pervading space thought to carry waves of light and vortices of matter in the manner that an ocean carries waves and vortices.*
Feynman diagram	*a diagram illustrating the exchange of virtual particles enacting a force between real particles.*
Fermions	*are subatomic particles defined by half Planck's constant, ½h. In the vortex theory they are considered to be particles formed of a single string of energy; one line of the motion of light.*
Gauge theory	*a theory in quantum mechanics which describes the assumed ability of force carrying virtual particles to feel or gauge the presence of other force carrying particles.*
Gamma-rays	*are high frequency photons emitted by radioactive atoms, blackholes and matter-antimatter annihilation.*
General Principle of Uncertainty	*The interpretations of scientific experiments cannot be certain if they are biased by the beliefs of those who interpret them.*
Gluons	*hypothetical force carrying particles that bind quarks in subatomic particles.*
Gravity	*an accelerating force of attraction that acts from the centre of every body of matter on every other body of matter.*
Gravity geyser	*a model to describe the action of a blackhole to emit energy in the manner of a geyser, which explains the periodicity of gamma ray bursts in distant galaxies.*
Heisenberg uncertainty principle	*proposes it is impossible to be simultaneously certain of the position and momentum of a primary particle of matter.*
Helium	*an element with two protons in its nucleus, it is the lightest inert gas.*
Higgs boson	*a particle thought to be responsible for mass.*
Isotope	*atoms of the same element with the same number of protons but a different number of neutrons.*
Kinetic energy	*the energy of motion usually associated with moving bodies or particles of matter.*
Magnetic moment	*a magnetic field generated by a spinning or rotating charged particle.*

Magnetism	*a force that is generated by the spin or rotation of charged particles or by an electric current.*
Mass	*In the vortex theory mass is defined as the ability of subatomic particles to resist movement and mass is considered to be conferred on them by the vortex spin of the energy forming them. In the standard theory in physics mass is defined in terms of Higgs scalar field which has two neutral and two electrically charged components that form a complex doublet of the weak isospin symmetry. Its sombrero potential leads it to a nonzero value everywhere - including space - which breaks the weak isospin symmetry of the electroweak interaction and, via the Higgs boson mechanism, confers mass on all massive elementary particles, including the Higgs boson itself.*
Materialism	*a theory, attributed to Democritus, that the universe is formed of atoms, indestructible particles of material substance that move about forever in a void of space. Also known as the atomic hypothesis.*
Neutron	*a subatomic particle that has a neutral charge and is slightly more massive than a proton that contributes binding energy without charge repulsion to the nucleus of an atom.*
Neutrino	*a neutral subatomic particle with miniscule mass and minimal interaction which has earned it the title of ghost or poltergeist particle. They are believed to be a fundamental, stable particle, the most common in the Universe with 100 trillion passing through our body every second. In the vortex theory the neutrino is considered to be unstable, a fragment of transient mass, a small amount of vortex energy escaping complete capture when a neutron decays into a proton and electron. The idea that they are the most abundant particle in the Universe and that trillions pass through us every second is considered to be conjecture.*
New Physics	*the standard model in theoretical physics based on quantum mechanics and quark theory.*
Nobel Prize	*five annual prizes - physics, chemistry, medicine, literature and peace - established by Alfred Nobel in 1895, consisting of a gold plate medal, a diploma and eleven million Swedish crowns.*
Nuclear fission	*the ability of heavy nuclei of uranium or plutonium to be split with neutrons to release nuclear energy.*
Nuclear fusion	*the ability of protons to fuse into atomic nuclei after high energy collisions with the release of nuclear energy.*
Nucleon	*a generic term for protons or neutrons in the nucleus of an atom.*
Occam's razor	*a thesis asserting that a simple theory which explains more things with less assumptions is more likely to be true than a complex theory requiring more assumptions to explain less things.*
Physics	*the study of physical systems.*
Photon	*a quantum of radiant energy in the visible spectrum. The term is now commonly attributed to radiant quanta of any frequency beyond and including the visible spectrum.*

Photon emission	*is photons released from an atom as electrons drop back from a higher level (excited state) orbit into a lower level (ground state) orbit after a quantum leap where the frequency of the emitted photon, reflected in its colour, is equivalent to the energy difference between the excited state and the ground state.*
Pion	*a pi or primary meson subatomic particle with neutral charge, a mass in the order of 14% of that of a proton and a lifespan in the order of trillionths of a second.*
Planck's Constant	*the constant defining a quantum of energy h named after Max Planck who initiated quantum theory. It is the constant of proportionality between energy E and frequency f in a quantum, i.e. $E=hf$.*
Potential energy	*the energy stored in bodies of matter mass, electrical energy, as nuclear energy or gravitational energy.*
Proton	*a heavy subatomic particle with a positive charge and life span estimated in quadtrillions of years.*
Quantum	*a particle of radiant energy governed by Planck's constant (h) which is sometimes called a photon.*
Quantum Electro Dynamics (QED)	*a theory in quantum mechanics which assumes electromagnetic forces are caused by the exchange of virtual photons between charged particles.*
Quantum leap	*the leap of an electron from a lower level (ground state) orbit in an atom into a higher level (excited state) orbit.*
Quantum loop symmetry	*the circulation of energy in the quantum and in the Universe.*
Quantum mechanics	*defines the properties and behaviour of particles of energy.*
Quantum spin	*the 3D spin of energy forming a subatomic particle.*
Quantum states	*energy states in an atom occupied by orbiting electrons. These include the higher level excited states and lower level ground states.*
Quantum theory	*established by Albert Einstein is the theory that energy is particulate.*
Quantum vortex theory	*the theory that subatomic particles are vortices of energy.*
Quark theory	*is the theory that subatomic particles are formed of a combination of hypothetical quarks: up-quark, down-quark, charmed-quark, strange-quark, top-quark and bottom quark. Quarks are supposed to have fractional electric charge.*
Quasars	*quasi-stellar formations that emit 200x the energy of a star, often associated with a blackhole.*
Radioactivity	*the property of some atoms to release radiation including helium nuclei, (alpha-particles) high speed electrons (beta-particles) and high frequency photons (gamma rays).*
Real Particles	*are particles that are both detectable and durable.*
SLAC	*the Stanford Linear Accelerator.*

Spectral lines	*the colour signature of heated atoms caused by photon emissions.*
Spinning string theory	*is the theory that particles are made up of vibrating or spinning strings of energy and that vibrating strings are fundamental to light and spinning strings are fundamental to matter and antimatter. The new vortex theory is a spinning string theory that preceded the generally accepted string theories that are based on vibration alone.*
String theory	*Contends that all particles are made up of vibrating strings or strands of energy.*
Strong nuclear force	*is caused by the exchange of pions between protons and neutrons which binds them together in the nucleus of an atom.*
The Thompson Principle	*to be real a particle must be both durable and detectable.*
Uncertainty principle	*also known as the indeterminacy principle, devised by Werner Heisenberg contends that it is not possible to determine both the position and momentum of a subatomic particle with any degree of accuracy.*
Virtual particles	*are neither detectable nor durable. They exist as mathematical constructs that serve theories in science.*
Virtual photons	*physically unreal particles in quantum mechanics that cannot be seen or detected that are thought to be responsible for electric and magnetic forces.*
Vortex atom	*a theory in nineteenth century physics that atoms were vortices in the ether.*
Vortex theory	*the theory that particles of energy can exist in the form of a vortex as well as a wave, that the vortex is fundamental to matter and is responsible for mass, static inertia, space, three dimensional extension, and force field interactions such as electric charge, magnetism and gravity are a consequence of vortex interactions.*
Wave-particle duality	*the property of primary particles of matter and light to behave simultaneously as waves and particles.*
Weak interaction	*also known as the weak nuclear force is the part of quark theory that allows quarks to swap their flavours i.e. an up quark into a down quark in the transformation of a proton and electron into a neutron.*

Index

V

W

X

Y

Z

About the Artist

As a child, I was captivated by art books showcasing the works of the old masters. To me, these images were a form of magic—these artists had the power to transform mere pigments into visions of other worlds. The art was beautiful, and mysterious, and beckoned me to be part of it. By the age of five, I had decided that I wanted to be a magician. It never crossed my mind that these creators were artists. It wasn't until my teenage years that I understood they were alchemists, turning crushed minerals mixed with linseed oil into visions that captured the ideologies—both real and imagined—of entire civilizations.

As I delved deeper, I realized that this was a key to a door that opened my mind to the invisible world, where mythologies explained the mysteries of the universe. Books like The Tao of Physics, The Vedic Texts, and The Egyptian Book of the Dead, written millennia ago, hinted at other realms that science would eventually begin to understand, bringing us to the present. The idea of God, as presented in Sunday School, was so abstract that it would overwhelm my young mind. But later, when I encountered concepts like the nonexistence of space and time at the quantum level and the idea that energy is constant and never diminishes, I realized that ancient texts were speaking of the same phenomena.

This revelation transformed my understanding of art. The works of the old masters, along with the art of ancient Egypt and the Mayans, transcended into an abstract realm of consciousness. That was my breakthrough as an artist: how could I paint the invisible and connect it to the history of Western painting? What is the concept of space within a picture plane? What if our conscious awareness is what illuminates what we see, and how could I depict that?

In June 2024, I was having an art opening in Paris when David Ash, slightly out of breath from rushing in from the rain, approached me in the gallery. Accompanied by his wife, he introduced himself and explained that he had come from England to meet me. When I asked why, he told me he was a quantum physicist writing a book, and what he was writing was what I was painting.

In that moment, the invisible threads of energy and thought connected us across the unseen realm, bringing us together. I think I'll call that magic.

- Shane Guffogg

About the Author

David Ash was born in Kent, England in 1948. His father, Dr Michael Ash, researching the link between cancer and radioactivity, introduced him to radioactivity physics when he was seven.

He discovered the vortex theory for the atom in 1965, during his last year at school, and began to collect ideas and data on subatomic matter in 1967 at the Hastings college of further education. David started work on the vortex theory at Queens University of Belfast in 1968, after Professor Gareth Owen, Dean of the Faculty of Science, encouraged him to use his time at university to develop a theory rather than cram for a career.

David went on to read physics and work on the vortex theory at Queen Elizabeth College, London University where he graduated with a Bachelor of Science degree in 1972. After a gap year he went onto the College of St Mark and St John in Plymouth in 1973 where he secured a postgrad in physics education in 1974.

While there he was introduced to the Royal Institution and joined it as a young member. On January 15th 1975 he gave a lecture on his vortex theory to young members and guests, from the rostrum where Lord Kelvin had demonstrated his vortex theory for the atom a century before.

David left London in 1976 to join his father who had founded CAMS, the College of Alternative Medicine and Science. There he taught the vortex theory, which was then published in part (with Peter Hewitt) in The Vortex: Key to Future Science, by Gateway books in 1990. After that David set out on an international lecture tour between 1991 and 1994.

In 1995 the College of Psychic Studies published his vortex theory in full as The New Science of the Spirit, which included a prediction of the accelerating expansion of the Universe. He has written and published 11 books, to date.

David lives in the UK and is father to nine and has sixteen grandchildren.

Mill House Publishers
Mill House, Kældehavegaard
Dæarupvej, Glamsbjerg
5620 Denmark
Reg: DK32589812

9788794302852

BES - #0019 - 290125 - C14 - 244/170/8 - CC - 9788794302852 - R - HT - Gloss Lamination